£30

CD

$$[\theta] = (\theta/(lc)) \times (M/100) = 2.30259 \times (4500/\pi)\Delta\varepsilon = 3300\Delta\varepsilon$$

$[\theta]$: molar ellipticity, θ : observed ellipticity angle,

l : cell length in dm, M : molecular weight,

c : concentration in g(solute)/mL(solution),

$\Delta\varepsilon$: molar circular dichroism ($\Delta\varepsilon = \varepsilon_l - \varepsilon_r$; $L \cdot mol^{-1} \cdot cm^{-1}$),

"A Value" : difference in $\Delta\varepsilon$ values of the two extrema of split
CD curves (see Figure 1-4)

In practice, the next equation is useful;

$$\theta/33 = \Delta\varepsilon \times c \times l$$

θ : observed ellipticity angle, l : cell length in cm,

c : concentration in mol(solute)/L(solution)

ORD

$$[\alpha] = \alpha/(lc) \qquad\qquad [\phi] = [\alpha] \times (M/100)$$

$[\alpha]$: specific rotation, α : observed rotation angle,

c : concentration in g(solute)/mL(solution); in case of neat
liquid, "c" is density in g/mL,

l : cell length in dm, $[\phi]$: molar rotation,

M : molecular weight

Sodium D-Line Rotation

e.g., $[\alpha]_D$ +100.0° (c 1.00, EtOH)

c : concentration in g(solute)/100mL(solution)

DATE DUE FOR RETURN

12/10/07

EXCITO

RY

CIRCULAR DICHROIC SPECTROSCOPY
—— EXCITON COUPLING IN ORGANIC STEREOCHEMISTRY ——

Nobuyuki Harada

Tohoku University, Sendai, Japan

Koji Nakanishi

Columbia University, New York, U.S.A.
Suntory Institute for Bioorganic Research, Kyoto, Japan

University Science Books
Oxford University Press
1983

First published 1983 by
University Science Books, 20 Edgehill Road, Mill Valley,
CA 94941, U.S.A.
and
Oxford University Press, Walton Street, Oxford OX2 6DP

Oxford   London   Glasgow   Melbourne
Wellington   Kuala Lumpur   Singapore   Jakarta   Hong Kong   Tokyo   Delhi
Bombay   Calcutta   Karachi   Ibadan   Nairobi   Dar es Salaam   Cape Town

ISBN 0 19 855709 4 OUP edition
ISBN 0 935702 09 1 USB edition

Library of Congress Cataloguing Data
81-51270

PREFACE

Although optical rotational data, recorded in the familiar form of $[\alpha]_D$, is one of the oldest of the physical constants, chiroptical properties were not used as a modern structural tool until the 1950's when the methods of optical rotatory dispersion and circular dichroism were revived by C. Djerassi and L. Velluz/ M. Legrand, respectively. However, relative to other physical data, there exists a large psychological barrier in the daily use of chiroptical properties. One reason for this may be the large element of empiricism, or the unfilled gap between practical data and theory.

The chiral through-space interaction between two or more chromophores, the thesis of this monograph, is based on the coupled oscillator method, and together with the X-ray Bijvoet method, enables one to determine the absolute configurations of organic compounds without reference to authentic cases. The theory was developed in the 1930's by W. Kuhn (coupled oscillator theory) and J. G. Kirkwood (group polarizability theory), and subsequently extended by W. Moffitt, I. Tinoco, Jr., J. A. Schellman (biopolymers), S. F. Mason (dimeric alkaloid), B. Bosnich (inorganic complex) and others.

The application of the coupled oscillator method to various natural products, which we have termed "the exciton chirality method", is in most cases straightforward; it can also be used for a variety of other purposes besides absolute configurational studies. We have attempted to *illustrate* the application of this versatile method by quoting in many cases figures from our own measurements; many of the data are unpublished. In the last three chapters, the coupled oscillator theory is described in a language more understandable to organic chemists; however, an understanding of the theory is not necessary for the actual usage of the method.

Naturally, we would more than welcome any criticism regarding the handling and interpretation of data.

We acknowledge the following for permission to adapt published Figures.

Bulletin of the Chemical Society of Japan	Figure 2-4, 2-5
Journal of the American Chemical Society,	Figure 2-6, 3-29
Journal of the Chemical Society,	
Perkin Transaction 1,	Figure 3-35
The Chemical Society of Japan/Maruzen, Tokyo,	
1965: Jikken Kagaku Koza, Zoku, vol. 11,	
"Electronic Spectroscopy,"	Figure 2-3
American Medical Association, Chicago, 1970:	
D. W. Urry, "Spectroscopic Approaches	
to Biomolecular Conformation,"	Figure 8-1
Benjamin, New York, 1965: N. J. Turro,	
"Molecular Photochemistry,"	Figure 11-11

Most studies from our laboratories were supported by grants from the Ministry of Education, Matsunaga, Science Foundation, the Japan Chemical Association (to N.H.), and National Institutes of Health Grants CA 11572 and AI 10187 (to K. N.).

Sendai, Japan   *Nobuyuki Harada*
New York City   *Koji Nakanishi*

May 28th, 1982

TABLE OF CONTENTS

I. CHIRAL EXCITON COUPLING AND CIRCULAR DICHROIC SPECTRA

1-1. Introduction

Organic molecules with π-electron systems interact with the electromagnetic field of ultraviolet or visible light to absorb the resonance energy corresponding to the energy gap between the ground and excited states. The UV and visible absorption spectra[1] of a variety of π-electron chromophores have been extensively studied and utilized for the acquisition of chemical information; conjugated or isolated double or triple bonds, hetero-double bonds such as C=O or C=N-, and aromatics are familiar chromophores in organic chemistry.

Similarly, circular dichroic spectroscopy (CD spectra)[2] of optically active compounds is a powerful method for studying three-dimensional structures of organic molecules. Namely, the method provides information on the absolute configuration, conformation, reaction mechanism, etc.

The electronic absorption and circular dichroic spectra are mostly determined in the solution state, but occasionally in the gas phase as well. These spectra result from the interaction of the individual chromophore of each molecule with the electromagnetic field of light, the interaction with neighbouring molecules being negligible. For example, in the case of a solution of 10^{-4} mol/L concentration, which is suitable for UV determination of

1

compounds having $\varepsilon = 10^4$, the average intermolecular distance is longer than 100 Å ; i.e., greater than the size of common organic molecules so that solute-solute interactions can be neglected.   Moreover, since molecules in solution are tumbling and randomly oriented, the mutual interaction between two mole-cules which is approximated by dipole-dipole interaction is negligible.

On the other hand, let us consider the changes in UV and CD spectra that may be anticipated by chromophore-chromophore interaction when aggregate sys-tems, i.e., dimer, trimer, ..., oligomer, polymer, and molecular crystals interact with light.   For example, in the case of a molecule having two iden-tical chromophores (i and j) connected by σ-bonds in some orientation (Figure 1-1), the two chromophores are brought to an excited state by the same proba-bility.   Namely, the probability of the state of excited chromophore i is exactly equal to that of the state of excited chromophore j; therefore mixing of the two state gives the excited state of the whole system.   In other words, the excited state (exciton)[3] delocalizes between the two chromophores i and j, just as in the case of electron delocalization in an ethylene double bond.

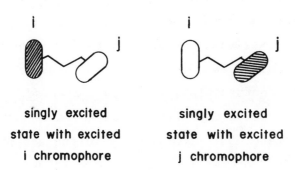

singly excited
state with excited
i chromophore

singly excited
state with excited
j chromophore

Figure 1-1.   _The excited state (exciton) delocalizes between two chromophores i and j._

The present interaction between excited states of chromophores — i.e., exciton coupling — was first studied in the field of electronic spectra of ionic or molecular crystals, and the theory of exciton coupling[4] was successfully applied to explain the so-called Davydov splitting in electronic absorption spectra.   In the field of optical activity of chiral compounds, Kuhn and Kirkwood[5] had investigated group-group interactions and proposed the coupled oscillator theory and the group polarizability theory, respectively, which have been extended to the chiral exciton coupling mechanism.

The exciton mechanism has been successfully employed in the studies of chiroptical properties of biopolymers,[6-8] (i.e., proteins, polypeptides, nucleic acids, oligonucleotides, dyes adsorbed on biopolymers, etc), and inorganic metal complexes;[9] it has also been the subject of theoretical treatments.[10,11]   Similarly, in the field of organic chemistry, circular dichroism due to the exciton coupling mechanism provides useful and unambiguous information on absolute configuration and conformation.[12-26, 38-46]   Since the chiral exciton coupling method is based on sound theoretical calculation, as will be shown in later chapters, the absolute stereochemistry of organic compounds exhibiting typical split CD Cotton effects due to chiral exciton coupling is assignable in a nonempirical manner.   The method has thus been extensively used for determining the absolute configuration of natural and synthetic organic compounds.

In the following section we will attempt to explain the mechanism of chiral exciton coupling in CD spectroscopy and the nonempirical nature of the method, and review recent developments of the present method in organic chemistry.

1-2. General Features of the Exciton Chirality Method in CD Spectroscopy and Definition of Exciton Chirality

General features and requirements of the exciton chirality method[18] in CD spectroscopy are as follows:

1. Two identical chromophores exhibiting strong $\pi \rightarrow \pi *$ absorption are located in chiral positions with respect to each other, as exemplified by the dibenzoate depicted in Figure 1-2.

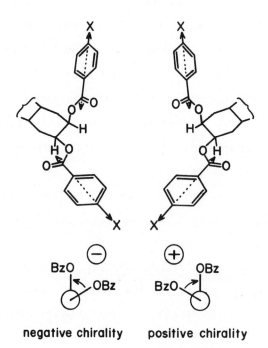

negative chirality   positive chirality

Figure 1-2. Chiralities of α-glycol dibenzoates. [Reprinted from reference 18.]

2. The exciton interaction between the two chromophores i and j splits the excited state into two energy levels, as shown in Figure 1-3. The energy gap $2V_{ij}$ is called the Davydov splitting.

3. Excitations to the two split energy levels generate Cotton effects of mutually opposite signs. As illustrated in Figure 1-4, this leads to a CD spectrum with two component Cotton effects of opposite signs which are separated by the energy gap $\Delta\lambda$ (Davydov splitting). Summation of the component Cotton effects results in the solid curve having two extrema. The extremum at longer and shorter wavelengths are called, respectively, the first Cotton effect and the second Cotton effect. In the case of ORD spectrum, three extrema are obtained by summation, as shown in Figure 1-4.

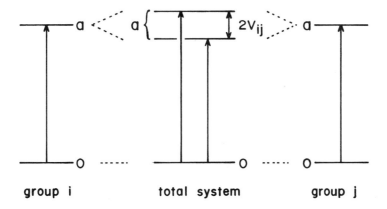

Figure 1-3. *By exciton interaction between the two chromophores i and j, excited state splits into two energy levels. The energy gap $2V_{ij}$ is called Davydov splitting.*

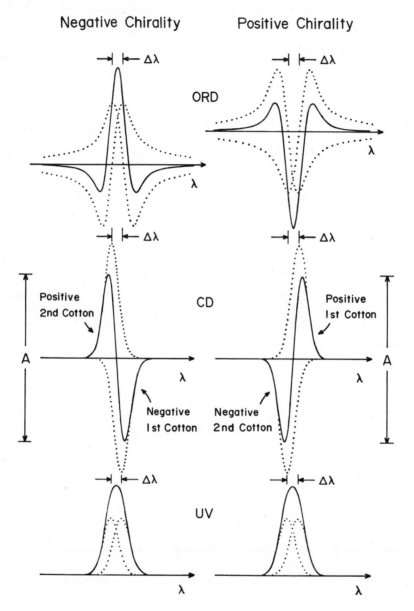

Figure 1-4. Summation curves (ORD and CD, solid lines) of two Cotton effects (broken lines) of opposite signs separated by Davydov splitting $\Delta\lambda$.   Positive

chirality gives positive first and negative second CD Cotton effects, while negative chirality gives negative first and positive second Cotton effects. The amplitude (A) of split CD Cotton effects is defined as: $A = \Delta\varepsilon_1 - \Delta\varepsilon_2$ where $\Delta\varepsilon_1$ and $\Delta\varepsilon_2$ are intensities of first and second Cotton effects, respectively. In the case of UV spectra, summation of the two component spectra leads to a single maximum. See Figures 1-6 and 1-9 for actual spectra.

4.   Provided the electric transition dipole moments of the two chromophores constitute a right-handed screwness (positive exciton chirality), as exemplified by the dibenzoate in Figure 1-2, the sign of the first Cotton effect is positive and that of the second Cotton effect is negative. When the two dipole moments make a left-handed screwness (negative exciton chirality), the signs of first and second Cotton effects are negative and positive, respectively (Figure 1-2 and Table 1-1).

5.   The exciton chirality governing the sign and amplitude of split Cotton effects is theoretically defined as:

$$\vec{R}_{ij} \cdot (\vec{\mu}_{i0a} \times \vec{\mu}_{j0a}) V_{ij}$$

where $\vec{R}_{ij}$ is interchromophoric distance vector from i to j, $\vec{\mu}_{i0a}$ and $\vec{\mu}_{j0a}$ are electric transition dipole moments of excitation $0 \to a$ of groups i and j, respectively, and V_{ij} is interaction energy between the two groups i and j (Table 1-1).

Table 1-1. Definition of Exciton Chirality for a Binary System.
[Reprinted from reference 21].

	Qualitative definition	Quantitative definition	Cotton effects
Positive chirality		$\vec{R}_{ij} \cdot (\vec{\mu}_{i0a} \times \vec{\mu}_{j0a}) V_{ij} > 0$	Positive first and negative second Cotton effects
Negative chirality		$\vec{R}_{ij} \cdot (\vec{\mu}_{i0a} \times \vec{\mu}_{j0a}) V_{ij} < 0$	Negative first and positive second Cotton effects

6.   If the direction of the transition moment in the chromophore — i.e.,
the polarization of the transition — is established, the signs of the
two Cotton effects enable one to determine the absolute stereochemistry of the two chromophores in space. Namely, the absolute configuration of a compound can be determined by the sign of Cotton effects on
the basis of the chiral exciton coupling mechanism.

7.   The present CD exciton chirality method is also applicable to compounds with three or more chromophores and compounds having chromophores that are not identical.

Typical examples of such chiral exciton coupling in CD spectra and its application to the configurational and conformational studies in organic chemistry are discussed in the following sections.

1-3. ORD, CD, and UV Spectra of Steroidal Dibenzoates

1-3-A. Cholest-5-ene-3β,4β-diol
Bis(p-dimethylaminobenzoate)[20,21]

The chromophore of p-dimethylaminobenzoate exhibits a strong $\pi \rightarrow \pi^*$ intramolecular charge transfer band around 310 nm, as shown in Figure 1-5. Namely, cholest-5-ene-3β-ol p-dimethylaminobenzoate (1) shows a $\pi \rightarrow \pi^*$ transition at 311.0 nm ($\varepsilon = 30,400$), together with a relatively weak transition at 229.0 nm ($\varepsilon = 7,200$). Similarly, cholest-5-ene-3β,4β-diol bis(p-dimethylaminobenzoate) (2) exhibits the same intramolecular charge transfer transition at 309.0 nm ($\varepsilon = 56,700$) and weak transition at 230.0 nm ($\varepsilon = 16,000$); the shape of the UV curve is quite similar to that of monobenzoate 1, but the molar extinction coefficient value is approximately twice because of the presence of two chromophores (see Figure 1-6). Thus, no remarkable change is observed between di- and mono-benzoate systems.

In contrast to the case of UV spectra, a striking difference is observed between CD spectra of mono- and di-benzoates. Namely, the CD spectrum of monobenzoate 1 exhibits a weak positive Cotton effect in the region of the intramolecular charge transfer band, $\lambda_{extremum}$* 309 nm, $\Delta\varepsilon$ +2.9, and a very weak positive one at 233 nm, $\Delta\varepsilon$ +0.7.

---

*$\lambda_{extremum}$:abbreviated to λ_{ext}.

On the other hand, as depicted in Figure 1-6, the CD spectrum of diben-zoate <u>2</u> exhibits two very strong Cotton effects of opposite signs in the region of the intramolecular charge transfer transition; the first Cotton effect at longer wavelength has a value of Δε −63.1 (320.5 nm), while the second Cotton effect at shorter wavelength has a value of Δε +39.7 (295.5 nm).

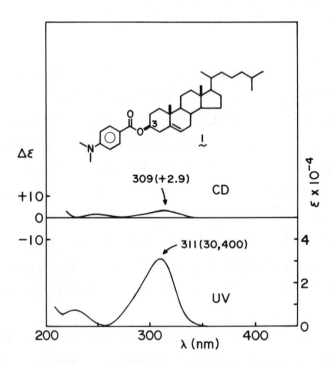

Figure 1-5. CD and UV spectra of cholest-5-ene-3β-ol p-dimethylaminobenzoate in 1% dioxane/EtOH.

Besides these strong Cotton effects, a relatively weak positive Cotton effect is observed at 227 nm ($\Delta\varepsilon$ +3.6), as in the case of monobenzoate 1. The summation of the amplitudes of the two Cotton effects gives the A-value = −102.8. Since the A-value is negative, a left-handed screwness exists between two transition dipole moments: i.e., the chirality is negative.

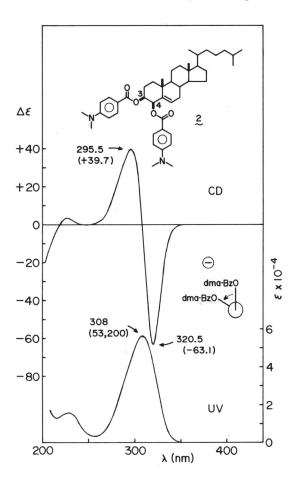

Figure 1-6.   CD and UV spectra of cholest-5-ene-3β,4β-diol bis(p-dimethylaminobenzoate) in EtOH.

Table 1-2.

Compound and Exciton Chirality	UV λ_{max},nm (ε)	CD λ_{ext},nm ($\Delta\varepsilon$)	References and Remarks
<u>2</u> ⊖	308 (53,200) 227 (13,400) EtOH	320.5 (-63.1) 308 (0.0) 295.5 (+39.7) EtOH	$\Delta\varepsilon_1/\Delta\varepsilon_2$ = -1.6 CD (20),(21) X ray (27)

As will be discussed in Chapter 10, the rotational strength of the Cotton
effect is proportional to the integrated peak area of Cotton effect plotted
against wavenumber. The ratio of the integrated peak area of the two observed
Cotton effects is (area of first Cotton effect): (area of second Cotton
effect) = 1.00 : 1.16. Thus, the rotational strengths of the two Cotton
effects are approximately equal to each other despite the imbalance between
the $\Delta\varepsilon$ values of the Cotton effects. Therefore, it is obvious that these
split type Cotton effects are due to the chiral exciton coupling between the
two dipole moments of the intramolecular charge transfer transition.

The intramolecular charge transfer transition of p–dimethylaminobenzoate
chromophore is polarized along the long axis of the chromophore, as illustrated in Figure 1-7. The present transition is mainly due to the charge transfer from the dimethylamino lone pair electrons to the carboxyl group via the
benzene moiety. Therefore, it is clear that the electric transition dipole
moment is along the long axis, which is nearly parallel to the alcoholic C-O

bond indicated by a heavy line in Figure 1-7.   This fact is of practical significance for the following reason: as previously mentioned, the exciton
chirality —— i.e., absolute geometry of two transition moments —— is assignable from the sign of the A-value of observed split type Cotton effects.   In
the case of the p-dimethylaminobenzoate chromophore, the exciton chirality
between the long axes of the two chromophores is approximated by the chirality
between two alcoholic C-O bonds.   In other words, the absolute geometry of two
alcoholic groups can be determined by observing the sign of the A-value of
split Cotton effects.

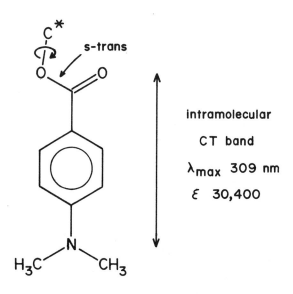

Figure 1-7. *Direction of the transition dipole moment in p-dimethylaminoben-*
zoate chromophore.

In the case of dibenzoate $\underline{2}$, the sign of the negative A-value leads to a left-handed screwness between the two hydroxyl groups, a conclusion which is in accordance with the results of X-ray crystallography,[27] i.e., the Bijvoet method, as depicted in Figure 1-8.

The present chiral exciton coupling method is based on nonempirical theoretical calculations, as will be discussed in Chapter 10. Therefore, the method enables one to determine the absolute stereochemistry in a nonempirical manner, as does the X-ray Bijvoet method. Namely this method provides a nonempirical rule for determining absolute configuration by means of CD spectroscopy. The results obtained by this method are more reliable than the empirical rules proposed in the field of organic chemistry.

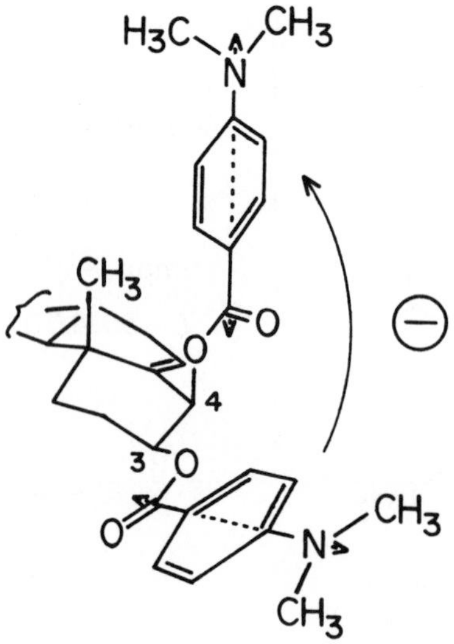

Figure 1-8. Negative exciton chirality of steroidal 3β,4β-dibenzoate system.

1-3-B.   5α-Cholestane-2α,3β-diol Dibenzoate[16]

The unsubstituted benzoate chromophore exhibits two UV absorption bands above 200 nm: at 280 nm ($\varepsilon \approx$ 1,000) due to ^1L_b transition of benzene ring and at 230 nm ($\varepsilon \approx$ 14,000) due to an intramolecular charge transfer or ^1L_a transition. The intense transition at 230 nm is polarized along the long axis of the chromophore because the transition originates from the charge transfer from benzene ring to ester group. On the other hand, the rather weak transition at 280 nm is polarized along the short axis of the chromophore.

As in the case of p-dimethylaminobenzoate, the intramolecular CT transition of unsubstituted benzoate is suitable for the CD exciton chirality method. For instance, 5α-cholestane-2α,3β-diol dibenzoate (3) exhibits the intramolecular CT transition at 229.3 nm (ε 26,700) in the UV spectrum. The CD spectrum of the dibenzoate 3 shows intense split Cotton effects of the exciton coupling type, i.e., negative first Cotton effect at 234 nm ($\Delta\varepsilon$ -13.9) and positive second Cotton effect at 219 nm ($\Delta\varepsilon$ +14.6). On the other hand, the ^1L_b transition around 280 nm only shows a weak single Cotton effect

Table 1-3.

(−)	229.3 (26,700)	234 (−13.9)	CD (16)
		228 (0.0)	
		219 (+14.6)	X ray (27)
3	10% dioxane/ EtOH	10% dioxane/ EtOH	

(Figure 1-9). The fact that this band is not split is mainly due to the weak intensity of this 280 nm ^1L_b transition. Thus, since Cotton effects due to exciton coupling are intense and appear as twin maxima of opposite signs, they are easily and clearly distinguished. The negative sign of the first

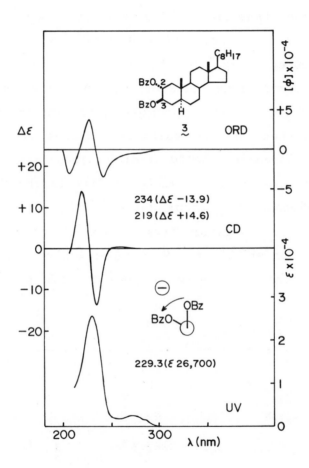

Figure 1-9. ORD, CD, and UV spectra of 5α-cholestane-2α,3β-diol dibenzoate in dioxane/EtOH.

Cotton effect leads to the conclusion that the exciton chirality between transition moments of two benzoate chromophores is negative and of left-handed screwness. Namely, the long axes of the two benzoate groups are twisted in a counterclockwise manner. This is consistent with the $2\alpha,3\beta$ configuration of the glycol moiety.

The ORD spectrum of $\underline{3}$ exhibits three extrema of negative/positive/negative signs from longer wavelength side, indicating a negative exciton chirality between the two benzoate groups. These ORD, CD, and UV spectra are quite similar in shape to the ideal curves of negative exciton chirality illustrated in Figure 1-4. Thus, absolute configuration of glycol system is easily determinable.

1-4.   CD and UV Spectra of (6R,15R)-(+)-6,15-Dihydro-6,15-ethanonaphtho[2,3-c]pentaphene[22-24]

Another typical example of chiral exciton coupling CD spectra is exemplified by (6R,15R)-(+)-6,15-dihydro-6,15-ethanonaphtho[2,3-c]pentaphene ($\underline{4}$) in this section.

Anthracene has four electronic transitions in the UV and visible region (Figure 1-10): an absorption band of medium intensity with complex vibrational structures at 356 nm (ε 7,600, ^1L_a transition), a very weak band buried in the ^1L_a band (^1L_b transition), a very intense absorption band at 252 nm (ε 204,000, ^1B_b transition), and an absorption of moderate intensity around 200 nm (ε 10,000, ^1C_b transition) (see section 2-4 for Platt's notation). Among these transitions, the ^1L_a and ^1B_b transitions are polarized along the short and long axes of the chromophore, respectively, as illustrated in Figure 1-10. In the case of the anthracene chromophore, it is clear that the ^1B_b transition is ideally suited for observing exciton coupling Cotton effects, because of the large molecular extinction coefficient values (ε value) and the established assignment of the polarization of the transition.

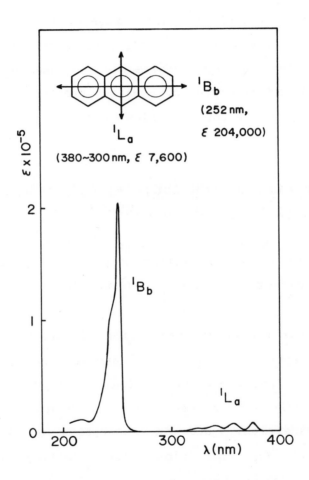

Figure 1-10.   *UV spectrum of anthracene in EtOH and polarization of the UV transitions.*

The electronic spectrum of hydrocarbon <u>4</u>, a cage molecule composed of two anthracene chromophores, resembles that of anthracene (Figure 1-11); the ^1L_a transition showing complex vibrational structures is located over 300–400 nm (371 nm, ε 11,200), and the allowed ^1B_b transition is at 267.2

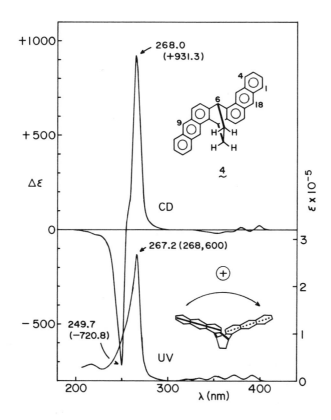

Figure 1-11. CD and UV spectra of (6R,15R)-(+)-6,15-dihydro-6,15-ethanonaph-tho[2,3-c]pentaphene.   Solvent: CD (0.18% dioxane/EtOH); UV (0.08% dioxane/EtOH). The positive sign of the first Cotton effect is in agreement with the positive exciton chirality between the long axes of the two anthracene moieties. [Adapted from reference 22.]

Table 1-4.

		397.2 (+26.4)	
	391.0 (9,100)	388.1 (-2.3)	
	371.2 (11,200)	378.0 (+6.3)	
		362.9 (-9.7)	CD (22)-(24)
	352.7 (9,000)	352.8 (-14.5)	
⊕	267.2 (268,600)	268.0 (+931.3)	Correl.
		256.0 (0.0)	(22),(23)
		249.7 (-720.8)	
4	0.08% dioxane/ EtOH	0.18% dioxane/ EtOH	

nm (ε 268,600). Thus, the whole pattern of the spectrum is quite similar to that of anthracene, which means that homoconjugation between the two anthracene chromophores is negligible.

The circular dichroic spectrum of compound 4 shows two extremely strong Cotton effects of opposite signs, a characteristic of the chiral exciton coupling mechanism, at the ^1B_b transition; the first Cotton effect located at longer wavelength has a $\Delta\varepsilon$ value of +931.3 (268.0 nm), while the second one at shorter wavelength has a $\Delta\varepsilon$ value of -720.8 (249.7nm): (A= +1652.1). These Cotton effects are the strongest in the field of organic compounds. Thus, as indicated by the quantitative definition of exciton chirality discussed in section 1-2, the allowed intense ^1B_b transition generates very intense Cotton effects because the exciton chirality is proportional to the square of transition dipole moment.

As in the case of dibenzoate 2, the peak areas of two Cotton effects are almost identical with each other; the ratio, (peak area of the first Cotton effect) : (peak area of the second one) is 1.00 : 1.01. This phenomenon proves that these two Cotton effects are exclusively due to the chiral exciton coupling mechanism.

Since the sign of the A-value is positive, the exciton chirality between two transition dipole moments of the ^1B_b transition is assigned to be positive. Namely, the chirality between the two long axes of the chromophores is of right-handed screwness (see Figure 1-11). Accordingly, the absolute configuration of the dextro-rotatory enantiomer (+)-4 is assigned to be (6R,15R).

1-5. CD Exciton Chirality Method and X-ray Bijvoet Method

The present CD exciton chirality method and the X-ray Bijvoet method[28,29] are nonempirical methods for determining absolute configuration of chemical compounds. Although the two methods are independent of each other, the conclusions obtained by both methods should naturally be consistent. In the case of compound 4, this was confirmed as follows. As summarized in Figure 1-12, compound 4 was synthesized[22,23] from diester (+)-5. The absolute configuration of compound (+)-5 had been established by Nakagawa, et al.[30] by means of chemical correlations with compounds (-)-6 and (+)-7, the absolute configuration of which in turn had been determined by the X-ray Bijvoet method.[31]   Thus, the conclusions obtained by both methods are in complete accord.[22,23]   Therefore, the CD data and synthesis of compound (+)-4 provide conclusive evidence demonstrating the consistency between the two nonempirical CD exciton chirality and X-ray Bijvoet methods.

Tanaka, et al.[32] had once claimed that the results of these two methods for determining absolute configurations were at variance, and that the CD method provided the correct configuration while the X-ray Bijvoet method should be revised.   However, this claim has since been retracted (private communication).   Mason[33] and Saito[34] have also independently pointed out the error in the theoretical CD and X-ray treatments of Tanaka.   Moreover, Brongersma and Mul[35] have experimentally proven that the X-ray Bijvoet method is correct.

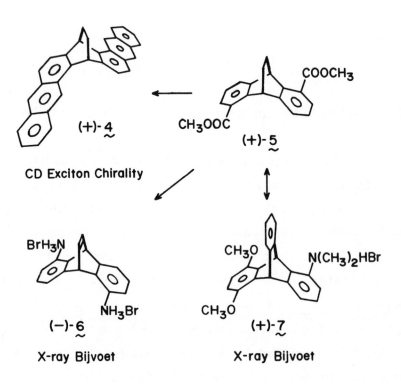

Figure 1-12.   Correlations between absolute stereochemistries of 9,10-dihydro-9,10-ethanoanthracene derivatives.   [Adapted from reference 23.]

1-6.   Development of the Coupled Oscillator Theory and CD Spectra of Calycanthine

The exciton chirality method originates from the coupled oscillator theory studied by Kuhn[5] and the polarizability theory by Kirkwood.[5]   The basic idea of the coupled oscillator theory has since been developed in the field of biopolymers and inorganic complexes,[9] and it has also been the subject of theoretical studies.[10,11]   For example, Moffitt[6] extended the method to study the chiroptical properties of biopolymers with identical groups. Subsequently, Tinoco[7] applied the coupled oscillator theory to polynucleotides, including di- and oligo-nucleotides; and Schellman[8] applied it to helical polypeptides. In the field of natural product chemistry, however, the exciton split Cotton effects have not been used for general structural studies of natural products except for the next isolated case.

Mason, et al.,[12] first applied the coupled oscillator theory to the CD spectrum of a natural product, calycanthine (8), a dimeric alkaloid with

Table 1-5.

		310 (6,300)	317 (+29)	
			289 (−5)	
			259 (+18)	CD (12)
	⊕	*252 (18,600)*		
			240 (−20)	X ray (34)
		EtOH	EtOH	

8

C_2-symmetry, to determine the absolute configuration. The UV spectrum of the aniline chromophore exhibits a benzenoid ^1L_a transition around 250 nm, which is polarized along the long axis of the chromophore. In calycanthine, a dimeric system composed of two aniline chromophores, the transition is split into two states, thus leading to a CD spectrum with split Cotton effects of opposite sign. The absolute configuration <u>8</u> depicted in Table 1-5 was assigned to calycanthine from the sign of the Cotton effects. Later, the configuration was verified by X-ray crystallographic studies.[36]

In principle, the concept of oscillator coupling is applicable not only to C_2-symmetrical compounds but also to other systems having two or more chromophores. In fact, as an extension of the benzoate sector rule,[37] an empirical rule for deducing absolute configurations of sec-hydroxyl groups, we have developed the dibenzoate chirality rule[16] for determining the absolute stereochemistry of glycol systems, most of which are not C_2-symmetrical. Further extension of this rule led to the exciton chirality method[18] which is widely applicable to interactions between different chromophores;[16-26] e.g., aromatics, conjugated polyenes, enones, esters, lactones, etc. The CD exciton chirality method thus has general applicability to a variety of complex natural products.

References

1. For monographs on UV spectroscopy see:

    (a) H. H. Jaffé and M. Orchin, <u>Theory and Applications of Ultraviolet Spectroscopy</u> (New York: Wiley, 1962).

    (b) A. I. Scott, <u>Interpretation of the Ultraviolet Spectra of Natural Products</u> (Oxford: Pergamon, 1964).

2. For monographs on CD and ORD spectroscopy see:

    (a) C. Djerassi, <u>Optical Rotatory Dispersion: Applications to Organic Chemistry</u> (New York: McGraw-Hill, 1960).

    (b) L. Velluz, M. Legrand, and M. Grosjean, <u>Optical Circular Dichroism: Principles, Measurements, and Applications</u> (Weinheim: Verlag Chemie, 1965).

    (c) P. Crabbé, <u>Optical Rotatory Dispersion and Circular Dichroism in Organic Chemistry</u> (San Francisco: Holden-Day, 1965).

    (d) G. Snatzke, ed. <u>Optical Rotatory Dispersion and Circular Dichroism in Organic Chemistry</u> (London: Heyden, 1967).

    (e) D. W. Urry, ed., <u>Spectroscopic Approaches to Biomolecular Conformation</u> (Chicago: American Medical Association, 1970).

    (f) D. J. Caldwell and H. Eyring, <u>The Theory of Optical Activity</u> (New York: Wiley-Interscience, 1971).

    (g) P. Crabbé, <u>Optical Rotatory Dispersion and Circular Dichroism in Chemistry and Biochemistry</u> (New York: Academic Press, 1972).

    (h) F. Ciardelli and P. Salvadori, eds., <u>Fundamental Aspects and Recent Developments in Optical Rotatory Dispersion and Circular Dichroism</u> (London: Heyden, 1973).

(i)   M. Legrand and M. J. Rougier, Stereochemistry: Fundamentals and Methods, ed. H. B. Kagan, (Stuttgart: Georg Thieme, 1977), Vo. 2.

(j)   S. F. Mason ed., Optical Activity and Chiral Discrimination (Dordrecht: D. Reidel, 1979).

(k)   E. Charney, The Molecular Basis of Optical Activity, Optical Rotatory Dispersion and Circular Dichroism (New York: Wiley, 1979).

(l)   P. Bayley, in An Introduction to Spectroscopy for Biochemists, ed. S. B. Brown (London: Academic Press, 1980), Chapter 5.

(m)   S. F. Mason, Molecular Optical Activity and the Chiral Discriminations (in press).

3.   J. I. Frenkel, Phys. Rev. 37, 17 (1931).

4.   A. S. Davydov, Zhur. Eksptl. i Teoret. Fiz. 18, 210 (1948) [Chem. Abstr. 43, 4575f (1949)]. A. S. Davydov, "Theory of Molecular Excitons," Trans. M. Kasha and M. Oppenheimer, Jr. (New York: McGraw-Hill, 1962).

5.   W. Kuhn, Trans. Faraday Soc. 26, 293 (1930). J. G. Kirkwood, J. Chem. Phys. 5, 479 (1937).

6.   W. Moffitt, J. Chem. Phys. 25, 467 (1956). W. Moffitt, D. D. Fitts and J. G. Kirkwood, Proc. Natl. Acad. Sci., U.S.A. 43, 723 (1957).

7.   I. Tinoco, Jr., Advan. Chem. Phys. 4, 113 (1962). I. Tinoco, Jr., R. W. Woody and D. F. Bradley J. Chem. Phys. 38, 1317 (1963). I. Tinoco, Jr., Radiat. Res. 20, 133 (1963). I. Tinoco, Jr. and C. A. Bush, Biopolym. Symp. 1, 235 (1964). W. C. Johnson and I. Tinoco, Jr., Biopolymers 8, 715 (1969).

8. J. A. Schellman and P. Oriel, *J. Chem. Phys.* 37, 2114 (1962). P. M. Bayley, E. B. Nielsen, and J. A. Schellman, *J. Phys. Chem.* 73, 228 (1969).

9. B. Bosnich, *Acc. Chem. Res.* 2, 266 (1969).

10. J. A. Schellman, *Acc. Chem. Res.* 1, 144 (1968).

11. A. D. Buckingham and P. J. Stiles, *Acc. Chem. Res.* 7, 258 (1974).

12. S. F. Mason, *Proc. Chem. Soc.* 1962, p. 362. S. F. Mason and G. W. Vane, *J. Chem. Soc. B* 1966, p. 370.

13. R. Grinter and S. F. Mason, *Trans. Faraday Soc.* 60, 274 (1964). S. F. Mason, K. Schofield, R. J. Wells, J. S. Whitehurst, and G. W. Vane, *Tetrahedron Lett.* 1967, p. 137. G. Gottarelli, S. F. Mason, and G. Torre, *J. Chem. Soc. B* 1970, p. 1349.

14. L. S. Forster, A. Moscowitz, J. G. Berger, and K. Mislow, *J. Am. Chem. Soc.* 84, 4353 (1962).

15. O. E. Weigang, Jr. and M. J. Nugent, *J. Am. Chem. Soc.* 91, 4555, 4556 (1969).

16. N. Harada and K. Nakanishi, *J. Am. Chem. Soc.* 91, 3989 (1969).

17. N. Harada, K. Nakanishi, and S. Tatsuoka, *J. Am. Chem. Soc.* 91, 5896 (1969). M. Koreeda, N. Harada, and K. Nakanishi, *Chem. Commun.* 1969, P.

548.   N. Harada and K. Nakanishi, Chem. Commun. 1970, p. 310.   N. Harada, H. Sato, and K. Nakanishi. Chem. Commun. 1970, p. 1691.   S. Marumo, N. Harada, K. Nakanishi, and T. Nishida, Chem. Commun. 1970, p. 1693.   T. Ito, N. Harada, and K. Nakanishi, Agr. Biol. Chem. (Japan) 35, 797 (1971).   N. Harada, S. Suzuki, H. Uda, and K. Nakanishi, J. Am. Chem. Soc. 93, 5577 (1971).

18.   N. Harada and K. Nakanishi, Acc. Chem. Res. 5, 257 (1972).

19.   N. Harada, J. Am. Chem. Soc. 95, 240 (1973).   M. Koreeda, N. Harada, and K. Nakanishi, J. Am. Chem. Soc. 96, 266 (1974).

20.   S.-M. L. Chen, N. Harada, and K. Nakanishi, J. Am. Chem. Soc. 96, 7352 (1974).

21.   N. Harada, S.-M. L. Chen, and K. Nakanishi, J. Am. Chem. Soc. 97, 5345 (1975).

22.   N. Harada, Y. Takuma, and H. Uda, J. Am. Chem. Soc. 98, 5408 (1976).

23.   N. Harada, Y. Takuma, and H. Uda, Bull. Chem. Soc. Jpn. 50, 2033 (1977).

24.   N. Harada, Y. Takuma, and H. Uda, J. Am. Chem. Soc. 100, 4029 (1978).

25.   N. Harada, N. Ochiai, K. Takada, and H. Uda, J. Chem. Soc., Chem. Commun. 1977, p. 495.   N. Harada, Y. Takuma, and H. Uda, Bull. Chem. Soc. Jpn. 51, 265 (1978).

26. N. Harada and H. Uda, J. Am. Chem. Soc. 100, 8022 (1978). N. Harada, Y. Tamai, Y. Takuma, and H. Uda, J. Am. Chem. Soc. 102, 501 (1980). N. Harada, Y. Tamai, and H. Uda, J. Am. Chem. Soc. 102, 506 (1980).

27. Absolute configurations of steroidal compounds are established by the X-ray Bijvoet method; see W. Klyne and J. Buckingham, Atlas of Stereochemistry (London: Chapman and Hall, 1978), vol. 1, pp. 121-26, vol. 2, pp. 63-64.

28. J. M. Bijvoet, A. F. Peerdeman, and A. J. Van Bommel, Nature 168, 271 (1951). J. M. Bijvoet, A. F. Peerdeman, and A. J. Van Bommel, Proc. K. Ned. Acad. Wet. B 54, 16 (1951). A. J. Van Bommel, Proc. K. Ned. Acad. Wet. B 56, 268 (1953). J. Trommel and J. M. Bijvoet, Acta Crystallogr. 7, 703 (1954). J. M. Bijvoet, and A. F. Peerdeman, Acta Crystallogr. 9, 1012 (1956).

29. See also J. Ibers and W. C. Hamilton, Acta Crystallogr. 17, 781 (1964).

30. Y. Shimizu, H. Tatemitsu, F. Ogura, and M. Nakagawa, Chem. Commun. 1973, p. 22. H. Tatemitsu, F. Ogura, and M. Nakagawa, Bull. Chem. Soc. Jpn. 46, 915 (1973).

31. N. Sakabe, K. Sakabe, K. Ozeki-Minakata, and J. Tanaka, Acta Crystallogr., Sect B 28, 3441 (1972). J. Tanaka, C. Katayama, F. Ogura, H. Tatemitsu, and M. Nakagawa, Chem. Commun. 1973, p. 21.

32. J. Tanaka, F. Ogura, H. Kuritani, and M. Nakagawa, Chimia 26, 471 (1972). J. Tanaka, K. Ozeki-Minakata, F. Ogura, and M. Nakagawa, Nature (London), Phys. Sci. 241, 22 (1973).   J. Tanaka, K. Ozeki-Minakata, F. Ogura, and M. Nakagawa, Spectrochim. Acta, Part A 29, 239 (1973).

33. S. F. Mason, J. Chem. Soc., Chem. Commun. 1973, p. 239.

34. Y. Saito, Kagaku to Kogyo (Japan) 26, 128 (1973).

35. H. H. Brongersma and P. M. Mul, Chem. Phys. Lett. 19, 217 (1973).

36. A. F. Beecham, A. C. Harley, A. McL. Mathieson, and J. A. Lamberton, Nature 244, 30 (1973).

37. N. Harada, M. Ohashi, and K. Nakanishi, J. Am. Chem. Soc. 90, 7349 (1968). N. Harada and K. Nakanishi, J. Am. Chem. Soc. 90, 7351 (1968).

38. Allylic benzoate method: N. Harada, J. Iwabuchi, Y. Yokota, H. Uda, and K. Nakanishi, J. Am. Chem. Soc., 103, 5590 (1981).

39. Additivity: H.-W. Liu and K. Nakanishi, J. Am. Chem. Soc., 103, 5591 (1981); H.-W. Liu and K. Nakanishi, J. Am. Chem. Soc., 104, 1178 (1982).

40. Use of a thiobenzoate chromophore: J. Gawronski, K. Gawronska, and H. Wynberg, J. Chem. Soc., Chem. Commun., 1981, p. 307.

41. Application of the allylic benzoate method: Y. Naya , K. Yoshihara, T.
     Iwashita, H. Komura, K. Nakanishi, and Y. Hata, J. Am. Chem. Soc., 103,
     7009 (1981).

42. Flower pigments: T. Hoshino, U. Matsumoto, N. Harada, and T.   Goto,
     Tetrahedron Lett., 22, 3621 (1981).

43. Application to oligosaccharides: H.-W. Liu and K. Nakanishi, J. Am. Chem.
     Soc., 103, 7005 (1981).

44. Application of the allylic benzoate method: W. H. Rastetter, J. Adams,
     and J. Bordner, Tetrahedron Lett., 23, 1319 (1982).

45. Absolute stereochemistry of 2,2'-spirobi-indane-1,1'-diols: N. Harada,
     T. Ai, and H. Uda, J. Chem. Soc., Chem. Commun. 1982, p.232.

46. Absolute configuration of brevetoxin B, a red tide toxin: ·Y. Y. Lin,
     M. Risk, S. M. Ray, D. V. Engen, J. Clardy, J. Golik, J. C. James, and
     K. Nakanishi, J. Am. Chem. Soc. 103, 6773 (1981).

II. ELECTRONIC TRANSITIONS OF CHROMOPHORES SUITABLE FOR CD EXCITON CHIRALITY METHOD

2-1. General Aspects of Chromophores Suitable for Chiral Exciton Coupling Mechanism

As mentioned in section 1-2, the following criteria have to be satisfied by chromophores used in the CD exciton chirality method.

1. Chromophores should exhibit strong $\pi \to \pi^*$ transition bands.
2. Polarization of the transition should be established in the geometry of the chromophore. For this reason, chromophores of high symmetry are desirable.

Various chromophores have been used for determining the absolute configurations of organic compounds by the exciton chirality method. In the following sections the electronic properties of these chromophores, i.e., λ_{max}, ε value, and polarization, are discussed.

2-2.   Para-substituted Benzoate Chromophores Suitable for Determining Absolute Configuration of Glycol Systems[1]

As exemplified in section 1-3, the intramolecular charge transfer transition of the para-substituted benzoate chromophores is suitable for determining the absolute chirality between two alcoholic groups.   In Table 2-1, the UV data of various para-substituted benzoate groups which have been utilized for the exciton chirality method are tabulated.   The intramolecular charge transfer band of the benzoate chromophore undergoes a red shift when electron-donating or -withdrawing groups are substituted in the para-position.[2]   Namely, the extent of the red shift of the transition increases with an increase in the electron donating properties of the para-substituent: i.e., $H < CH_3 < Cl < Br < OCH_3 < NH_2 < N(CH_3)_2$.   The same is true for the case of electron-withdrawing groups: i.e., $H < CN < NO_2$ (see Figure 2-1).

The direction of the dipole moment of the intramolecular charge transfer transition is along the long axis of the chromophore in all cases listed in Table 2-1.   Namely, the transition consists almost entirely of an intramolecular charge transfer from the electron-donating substituent to the ester moiety through the benzene ring.   Therefore, it is understandable that the transition is polarized along the long axis of the chromophore.   Similarly, in the case of electron-withdrawing groups, charge transfer occurs from the benzene ring to the two para-substituent groups, thus causing the long-axis polarization. The transition dipole moments of the intramolecular charge transfer band of para-substituted benzoates are therefore parallel to the long axis, and hence to a first approximation, parallel to the alcoholic C-O bond; this enables one to determine the absolute chirality of glycol systems.

Table 2-1. **UV Spectral Data of** <u>Para</u>**-substituted Benzoates of Cholesterol.**

Chromophore	Intramolecular CT or ^1L_a transition	^1L_b and other transitions	Solvent
^1L_b CT or ^1L_a	229.5 nm ε 15,300	273.6 nm ε 900	EtOH
CH₃	238.4 nm ε 17,600	280.6 nm ε 600	EtOH
Cl	240.0 nm ε 21,400	282.5 nm ε 600	EtOH
Br	244.5 nm ε 19,500	283.0 nm ε 500	EtOH/ dioxane (280:1)

OCH$_3$ structure	257.0 nm ε 20,400	——^{a}	EtOH
NH$_2$ structure	293.8 nm ε 21,900	——^{a}	MeOH/ dioxane (9:1)
H$_3$C–N–CH$_3$ structure	311.0 nm ε 30,400	229.0 nm ε 7,200	EtOH
CN structure	240.0 nm ε 24,600	283.4 nm ε 1,700	EtOH

NO₂

260.5 nm

ε 15,100

————ᵃ

EtOH/
dioxane
(24:1)

ᵃ Buried in intense intramolecular charge transfer transition.

Of the various para-substituted benzoate chromophores, the p-dimethylami-nobenzoate group[3,4] is superior to others because of the greatest extent of the red shift and the largest ε value. The split CD Cotton effects due to interacting p-dimethylaminobenzoate groups are thus much more readily observable as compared to those arising from other benzoate transitions.

Based on the above discussion, it is easily understandable that ortho- or meta-substituted benzoates are not suited for the CD exciton chirality method. Namely, these chromophores have less symmetry, and moreover, the polarization of the intramolecular charge transfer transition is not parallel to the alcoholic C-O bond. The exciton chirality between the two chromophores therefore depends on the conformational rotation of the alcoholic C-O bonds, which in turn prevents determination of the absolute chirality between two alcoholic C-O bonds. The chiral exciton coupling method therefore requires chromophores of high symmetry and of strong absorption.

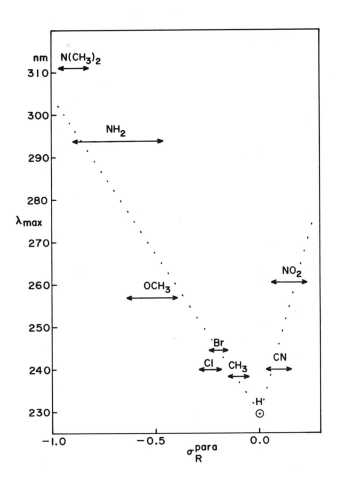

Figure 2-1. UV λ_{max} _of_ _para-substituted_ _benzoate_ _chromophores_ _plotted_ _against Taft's resonance parameter_ σ_R^{para}, _where_ ↔ _indicates_ _the_ _range_ _of_ σ_R^{para} _values._[5]

Rigidly speaking, the β-naphthoate chromophore cannot be used in spite of the fact that they lead to intense CD split Cotton effects.   However, as depicted in Figure 2-2, the ^1B_b band of cholest-5-ene-3β,4β-diol bis(β-naphthoate) (<u>1</u>) shows two split Cotton effects at 242 nm, Δε –214.5 and at 229 nm, Δε + 191.9, which are much stronger than those of the <u>p</u>-dimethylaminobenzo-

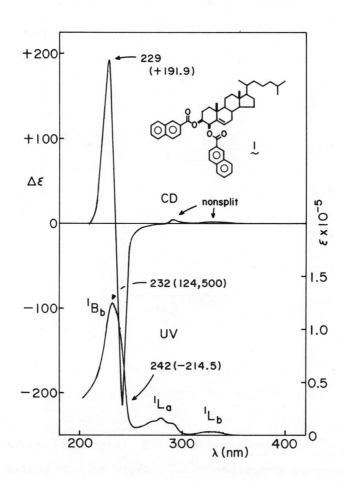

Figure 2-2. *CD and UV spectra of cholest-5-ene-3β,4β-diol bis(2-naphthoate) in dioxane/EtOH.*

Table 2-2.

334 (3,200)	334 (+1.7)		
	290 (+3.6)		
280 (16,000)	282 (-1.0)		
	242 (-214.5)	CD (1)	
232 (124,500)	*236 (0.0)*		
	229 (+191.9)	X ray (6)	
0.7% dioxane/ EtOH	10% dioxane/ EtOH		

ate[1] (see Figure 1-6), and the signs of which are in agreement with those expected from the helicity between the 3β- and 4β-C-O bonds. This is probably due to the fact that, unlike the o- and m-substituted benzoate chromophores, the ^1B_b band of the β-naphthoate group is approximately parallel to the C-O bond.

2-3. Benzamido Chromophore for Amino Alcohols and Diamines

The exciton chirality method is applicable to the intramolecular charge transfer band of benzamido groups as well as alcohol benzoates. As indicated in Table 2-3, the intramolecular charge transfer transition or ^1L_a transition of the benzamido group is located at 225 nm (ε 11,200) and is polarized along the long axis of the chromophore.

Table 2-3.   UV Data of N-cyclohexylbenzamide.

Chromophore	Intramolecular CT or ^1L_a transition	^1L_b and other transitions	Solvent
CT	224.6 nm $\mathcal{E}$ 11,200	—[a]	EtOH

[a]   Not observed.

Because of the proximity between the λ_{max} values of benzamido and benzoate chromophores, two chromophores couple with each other.   Therefore, the exciton chirality method is applicable to amino alcohol and diamine systems, as will be discussed in Chapter 3.

2-4.   Polyacene Chromophores for the Exciton Chirality Method

As described in section 1-4, the ^1B_b transition of the polyacene chromophore is ideally suited for the exciton chirality method.   Table 2-4 shows the UV spectral data of some polyacenes including the benzene chromophore.

In these polyacene chromophores, all chromophores are of D_{2h}-symmetry, except benzene which is of D_{6h}-symmetry.   Accordingly, the long and short axes of polyacene chromophores, i.e., the polarization of transitions, are definitely assignable.   On the other hand, the benzene chromophore has

Table 2-4.   UV Spectral Data of Polyacenes.

Chromophore	$^{1}B_b$ transition	$^{1}L_a$ and $^{1}L_b$ transitions	Solvent
$^{1}B_b$ $^{1}L_a$	220.2 nm $\mathcal{E}$ 107,300	275.5 nm[a] $\mathcal{E}$ 5,800 312.0 nm[b] $\mathcal{E}$ 200	EtOH
	251.9 nm $\mathcal{E}$ 204,000	356.5 nm[a] $\mathcal{E}$ 7,600	EtOH
	274.0 nm[c] $\mathcal{E}$ 316,000	471.0 nm[a] $\mathcal{E}$ 10,000	EtOH

a $^{1}L_a$ transition.   b $^{1}L_b$ transition.   c [reference 7].

three long and three short axes.   Therefore, the effect of weak substituents on the benzene ring — e.g., alkylated benzenes — is insufficient to assign the direction of polarization.   Thus, the alkylated benzene chromophores are, in general, unsuited for the exciton chirality method.

Appendix.   Platt's Notation and Polarization Diagram[8]

Q: quantum number of angular momentum of excited state.

n: number of benzene ring.

$$Q = \quad 0, \qquad \pm 1, \qquad \pm 2, \qquad \cdots\cdots\cdots$$
$$A, \qquad B, \qquad C, \qquad \cdots\cdots\cdots$$

$$Q = \pm(2n+1), \quad \pm(2n+2), \quad \pm(2n+3), \quad \cdots\cdots\cdots$$
$$L, \qquad M, \qquad N \qquad \cdots\cdots\cdots$$

Node of excited wavefunction is on atoms $\cdots\cdots\cdots$ b

Node is between atoms$\cdots\cdots\cdots\cdots\cdots\cdots\cdots\cdots\cdots\cdots$ a

The case of naphthalene is exemplified in Figure 2-3.

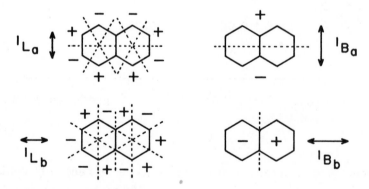

Figure 2-3. *Platt's polarization diagram of naphthalene.* *[Adapted from reference 9.]*

2-5.   Conjugated Olefins, Ketones, Esters, and Lactones for the Exciton Chirality Method

The $\pi \to \pi^*$ transitions of conjugated olefins, ketones, esters, and lactones are also useful for the exciton chirality method.[10]   As exemplified by the Woodward–Fieser empirical UV rules,[11,12] the electronic properties, — i.e., λ_{max}, ε value, and the direction of transition moment, — of conjugated dienes depend on the s-trans or s-cis conformation.   Typical examples of these systems are shown in Table 2-5.   In general, the absorption maximum of the

Table 2-5.   **UV Spectral Data of Typical Conjugated Dienes.**

Chromophore		$\pi - \pi^*$ transition	Solvent
s–cis		265 nm[a] ε 6,400	isooctane
s–trans		234 nm[b] ε 20,000	EtOH

[a]   Reference 13.   [b] Reference 12.

<u>Table 2-6.</u> **UV Spectral Data of Conjugated Enone, Esters, and Lactone.**

Chromophore	$\pi - \pi^*$ transition	Solvent
Steroid	241 nm[a,b] ε 16,600	EtOH
	215 nm[c] ε 11,200	EtOH
	217 nm[d] ε 15,100	MeOH
	259.4 nm ε 24,700	EtOH

[a] The CD spectra of enones clearly show the <u>presence of a third transition</u> <u>around 210 nm</u> in addition to the familiar $n \rightarrow \pi^*$ and $\pi \rightarrow \pi^*$ transitions:L. Velluz, M. Legrand, and R. Viennet, <u>Compt. Rend.</u> <u>261</u>, 1687 (1965).   A. W. Burgstahler and R. C. Barkhurst, <u>J. Am. Chem. Soc.</u> <u>92</u>, 7601 (1970).   The assignment of this transition is unknown.   [b] Reference 14.   [c] Reference 15.   [d] Reference 16.

s-<u>cis</u> form is located at longer wavelength than that of the s-<u>trans</u> form, while the absorption intensity ε of the s-<u>cis</u> form is weaker than that of the s-<u>trans</u> form.

Typical examples of other chromophores, conjugated ketones, esters, and lactones are also tabulated in Table 2-6.

2-6. Other Substituted Benzenoid Chromophores

Phenylacetylene and benzonitrile chromophores exhibit intramolecular charge transfer transitions around 230 nm, similar to the benzoate chromophore. As indicated in Table 2-7, the CT transitions are polarized exactly parallel to

Table 2-7. UV Spectral Data of Phenylacetylene and Benzonitrile.

Chromophore	Intramolecular CT or ^1L_a transition	^1L_b transition	Solvent
CT or ^1L_a	234.2 nm ε 15,000	269.5 nm ε 350	EtOH
	227.6 nm ε 14,200	284.0 nm ε 1,900	EtOH

the long axis of the chromophores because of their highly symmetrical structures.

2-7. Polarization of Electronic Transitions

The polarization of electronic transitions is experimentally and theoretically
determinable by employing the following methods:

   1.   Linear dichroic spectrum of single crystals.
   2.   Linear dichroic spectrum by the stretched film method.
   3.   Polarized luminescence spectrum.[17]
   4.   Molecular orbital calculation.

In this section, methods 1, 2, and 4 are briefly described.

Dichroic Spectrum of Benzoic Acid Single Crystal. The crystal structure
of benzoic acid has been determined by X-ray crystallography, and the optical
properties of the crystal itself have also been studied; these studies show
that the molecules are aligned in the crystal lattice as depicted in Figure
2-4. Namely, two benzoic acid molecules are paired by hydrogen bonding, and
the long axis of the molecule is almost along the b-axis, while the short axis
is along the a-axis. Tanaka[18] measured the dichroic spectrum of the
single crystal of benzoic acid to arrive at the result illustrated in Figure
2-5. The solid line is the absorption spectrum excited by incident light
polarized parallel to the a-axis, while the dotted curve is the spectrum with
incident light polarized parallel to the b-axis.

From the spectra, it is readily seen that the intense absorption around 230 nm is polarized almost parallel to the long axis of the chromophore, while the weak transition around 280 nm is polarized parallel to the short axis. The calculation of intensity ratio confirmed the assignment in a quantitative manner.   Thus, the long-axis polarization of the intramolecular CT or ^1L_a

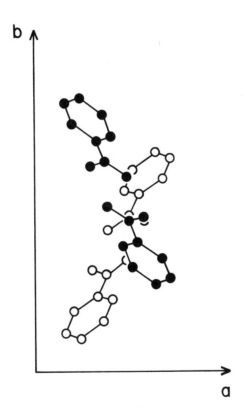

Figure 2-4.   *Projection of benzoic acid molecules onto the (001) plane in a single crystal.   [Adapted from reference 18.]*

transition (230 nm) of benzoic acid chromophore has been established.   By the use of this method, polarization properties of various organic molecules have been clarified.

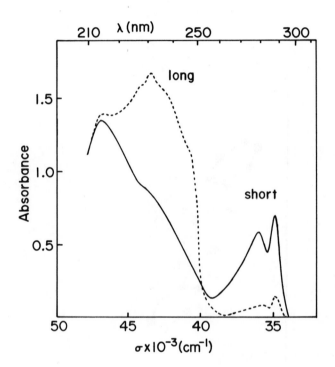

Figure 2-5.   *Polarized absorption spectrum of benzoic acid crystal: (⎯⎯⎯⎯) absorption spectrum excited by a light polarized parallel to the a-axis of Figure 2-4; (•••••) absorption spectrum excited by a light polarized parallel to the b-axis of Figure 2-4. [Adapted from reference 18.]*

Linear Dichroic Spectra of p-Methoxybenzoic Acid on a Stretched Polyethylene Film. Yogev and coworkers[19] measured the linear dichroic spectra of
p-methoxybenzoic acid by a stretched film method as follows: p-methoxybenzoic
acid was dissolved in chloroform, and a 5×2.5×0.01 mm polyethylene film was
dipped in the solution. After being kept at 60 °C for 24 h, the film was
washed, dried, and stretched. Upon stretching, the p-methoxybenzoic acid

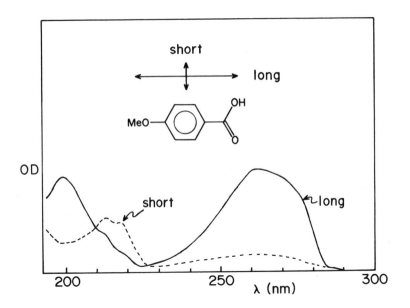

*Figure 2-6. Linear dichroic spectrum of p-methoxybenzoic acid on a poly-
ethylene film:(———) absorption spectrum excited by polarized light
parallel to the direction of stretch; (•••••) absorption spectrum excited by
polarized light perpendicular to the direction of the stretch. [Adapted from
reference 19].*

molecules incorporated in the film become partially aligned along the direction of stretch.   The tendency of the alignment is further enhanced by a
hydrogen-bonding association of two molecules.   The resultant dichroic spectra
are shown in Figure 2-6, in which the solid line is the absorption spectrum
excited by incident light polarized parallel to the direction of stretch.   The
dotted curve is the spectrum with incident light polarized perpendicular to
the direction of stretch.   Thus, the spectra clearly indicate that the 260 nm
transition, i.e., an intramolecular charge transfer transition, is parallel to
the long axis of the chromophore.   The present method is simple and applicable
to various compounds, e.g., anthracene[20,21] and steroidal enones.[22,23]

Molecular Orbital Calculation of the UV Spectra of Anthracene.   Polarization of electronic transitions is theoretically calculable by the molecular
orbital method.   For instance, in the case of anthracene, the calculation
results obtained by the Pariser-Parr-Pople method (Table 2-8) clearly indicate
that the ^1B_b transition around 250 nm is polarized parallel to the long
axis of the chromophore, while the ^1L_a transition around 360 nm is polarized along the short axis.   Thus, polarization of electronic transitions is
determinable by both experimental and theoretical methods.

Table 2-8. Observed and Calculated UV Spectra of Anthracene[24]

Observed (in EtOH)		Calculated (SCF-CI-DV)[a]		
λ_{max} (nm),(ε)	$D \times 10^{36}$	λ_{max} (nm)	$D \times 10^{36}$	Polarization
374.5 (7,500)				
356.5 (7,600)	2.70 ^1L_a	361.0	2.90	y-axis(short)
339.6 (5,300)				
324.0 (2,800)				
310.0 (1,200)				
251.9 (204,000)	- 87.89 ^1B_b	245.5	87.89	x-axis(long)

[a] Dipole strengths are in cgs unit.

References

1. N. Harada and K. Nakanishi, <u>Acc. Chem. Res.</u> <u>5</u>, 257 (1972).

2. N. Harada and K. Nakanishi, <u>J. Am. Chem. Soc.</u> <u>90,</u> 7351 (1968).

3. S.-M. L. Chen, N. Harada, and K. Nakanishi, <u>J. Am. Chem. Soc.</u> <u>96</u>, 7352 (1974).

4. N. Harada, S.-M. L. Chen, and K. Nakanishi, <u>J. Am. Chem. Soc.</u> <u>97</u>, 5345 (1975).

5. R. W. Taft and I. C. Lewis, <u>J. Am. Chem. Soc.</u> <u>81</u>, 5343 (1949).

6. Absolute configurations of steroidal compounds are established by the X-ray Bijvoet method; see W. Klyne and J. Buckingham, <u>Atlas of Stereo-chemistry</u> (London: Chapman and Hall, 1978), vol. 1, pp. 121-26, vol. 2, pp. 63-64.

7. E. Clar and C. Marshalk, <u>Bull. Soc. Chim. Fr.</u> <u>17</u>, 433 (1950).

8. J. R. Platt, <u>J. Chem. Phys.</u> <u>17</u>, 484 (1949).

9. S. Nagakura, J. Tanaka, K. Kimura, M. Ito, I. Hanasaki, K. Kaya, and A. Ishitani, <u>Jikken Kagaku Koza, Zoku</u>, vol. 11, "Electronic Spectra," ed. S. Nagakura, (Tokyo: Maruzen, 1965), p. 289.

10. M. Koreeda, N. Harada, and K. Nakanishi, <u>J. Am. Chem. Soc.</u> <u>96</u>, 266 (1974).

11. R. B. Woodward, J. Am. Chem. Soc. 64, 72 (1942).

12. L. F. Fieser and M. Fieser, Steroids (New York: Reinhold, 1959) p. 15.

13. R. T. O'Connor and L. A. Goldblatt, Anal. Chem. 26, 1726 (1954).

14. L. Dorfman, Chem. Rev. 53, 47 (1953).

15. S. M. Kupchan and A. Afonso, J. Org. Chem. 25, 2217 (1960).

16. R. Tschesche and H. Machleidt, Justus Liebigs Ann. Chem. 631, 61 (1960).

17. For example, R. Shimada and L. Goodman, J. Chem. Phys. 43, 2027 (1965).

18. J. Tanaka, Bull. Chem. Soc. Jpn. 36, 833 (1963).

19. A. Yogev, L. Margulies, and Y. Mazur, J. Am. Chem. Soc. 92, 6059 (1970).

20. H. Inoue, T. Hoshi, T. Masamoto, J. Shiraishi, and Y. Tanizaki, Ber. Bunsenges. Phys. Chem. 75, 441 (1971).

21. J. Michl, E. W. Thulstrup, and J. H. Eggers, Ber Bunsenges. Phys. Chem. 78, 575 (1974).

22. A. Yogev, L. Margulies, D. Amar, and Y. Mazur, J. Am. Chem. Soc. 91, 4558 (1969).

23. A. Yogev, J. Riboid, J. Marero, and Y. Mazur, <u>J. Am. Chem. Soc.</u> <u>91</u>, 4559 (1969).

24. N. Harada, unpublished.

25. Use of a thiobenzoate chromophore: J. Gawronski, K. Gawronska, and H. Wynberg, <u>J. Chem. Soc., Chem. Commun.,</u> 1981, p.307.

III. APPLICATION OF THE EXCITON CHIRALITY METHOD FOR DETERMINATION OF ABSOLUTE CONFIGURATION

3-1. Dibenzoate Chirality Method for Glycol Systems

3-1-A. Distant Effect of Split Type Cotton Effects; Bis(p-dimethylaminobenzoates) of Steroidal Glycols[1,2]

Typical examples of CD split Cotton effects due to chiral exciton coupling between two intramolecular charge transfer transitions of bis(p-dimethylami-nobenzoates) of steroidal glycols are shown in the following tables and figures. Figure 3-1 shows general features of UV and CD spectra of bis(p-dimethylaminobenzoate) systems. As expected, the sign and amplitude of split Cotton effects depend on the absolute disposition of two benzoate chromophores; i.e., the interchromophoric distance and angles. The position of the split Cotton effects, however, is fixed at specific wavelengths within a range of a few nm, irrespective of difference in the sign and amplitude of Cotton effects, interchromophoric distances, and angles. Namely, the first Cotton effect is located at 319-321 nm, while the second Cotton effect is at 291-295 nm. Moreover, the position of the zero line intersection flanked by two Cotton effects is almost fixed at 305-308 nm, which is close to the UV maximum position of 307-310 nm.

In addition, the ratio of $\Delta\varepsilon$ values of the first and second Cotton effects is almost constant, i.e., about 2 : 1 (see "Remarks" in Tables 3-1 through 3-4). Thus, the shapes of split Cotton effects of bis(p-dimethylami-nobenzoates) are similar to each other, except for the signs and amplitudes. Namely, as theoretically discussed in Chapters 10 through 12, the pattern of CD curves reflects that of UV absorption band (if the UV absorption band is

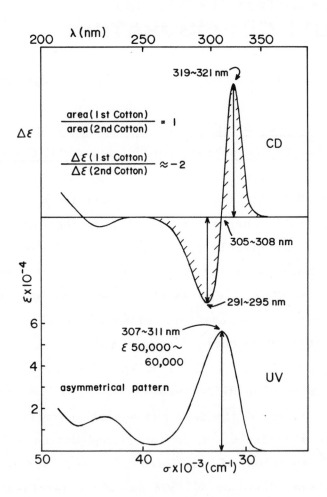

Figure 3-1. *General pattern of UV and CD spectra of bis (p-dimethylaminoben-zoate) systems.*

Table 3-1.

Compound and Exciton Chirality	UV λ_{max}, nm (ε)	CD λ_{ext}, nm ($\Delta\varepsilon$)	References and Remarks
	307 (54,300) EtOH	320 (+61.7) 307 (0.0) 295 (-33.2) EtOH	$\Delta\varepsilon_1/\Delta\varepsilon_2$ = -1.9 CD (1),(2) X ray (3)

approximated by a Gaussian distribution, the ratio of $\Delta\varepsilon$ values is 1 : 1).
These features are summarized in Figure 3-1. In view of these features of
split Cotton effects, it is quite easy to discriminate the CD Cotton effects
of bis(p-dimethylaminobenzoates).

In the case of 5α-cholestane-2β,3β-diol bis(p-dimethylaminobenzoate)
(1), the CD spectrum exhibits positive first and negative second Cotton ef-
fects, λ_{ext} 320 nm, $\Delta\varepsilon$ +61.7 and λ_{ext} 295 nm, $\Delta\varepsilon$ -33.2, in the region of the
$\pi\rightarrow\pi^*$ intramolecular charge transfer or ^1L_a transition, λ_{max} 307 nm, ε 54,300
(Table 3-1 and Figure 3-2). The sign of the exciton split Cotton effects is
in agreement with the positive exciton chirality between the 2β- and 3β-benzo-
ate chromophores. The ratio $\Delta\varepsilon_1/\Delta\varepsilon_2$ is -1.9 (Table 3-1); the first Cotton
effect is approximately twice as large as the second Cotton effect.

As exemplified by cholest-5-ene-3β,4β-diol bis(p-dimethylaminobenzoate)
in Chapter 1 and compounds 1 and 2, bis(p-dimethylaminobenzoates) of vicinal
glycols with a 60° dihedral angle generally exhibit intense split Cotton

effects because of the proximity of the two p-dimethylaminobenzoate chromophores (A value ≃ 100).   Especially, compound 3 shows the largest split Cotton effects with an A value of 140.   This may be due to sterical crowding of the two chromophores caused by ring B.   On the other hand, in the case of com-

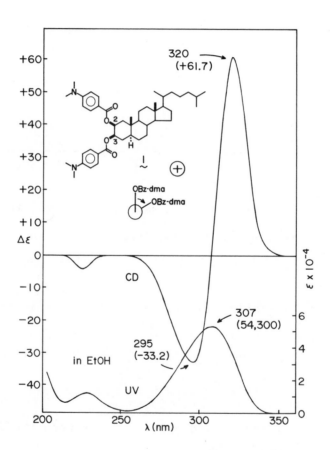

Figure 3-2. CD and UV spectra of 5α-cholestane-2β,3β-diol bis(p-dimethylamino-benzoate) in EtOH. [Adapted from reference 2.]

Table 3-2.

		$\Delta\varepsilon_1/\Delta\varepsilon_2$
		= -2.3
	321 (+61.3)	
308 (51,200)	308 (0.0)	CD (1),(2)
	295 (-27.1)	
		X ray (3)
EtOH	EtOH	

		$\Delta\varepsilon_1/\Delta\varepsilon_2$
	321.9 (+91.3)	
308.7 (52,900)	309.2 (0.0)	= -1.7
	297.4 (-52.5)	
	238.8 (+3.5)	CD (5)
226.7 (14,400)	225.5 (-3.1)	
	204.6 (+37.6)	X ray (3)
EtOH	EtOH	

---

pounds 1 and 2, the benzoate groups are flanked by methylene moieties and therefore can adopt a more remote disposition.

It should be also noted that the Δε values of compounds 1 and 2 are quite similar to each other, in spite of the difference in configurational stereo-chemistry. This phenomenon demonstrates that the split CD Cotton effects are mainly dependent on the mutual disposition between the two benzoate chromo-phores, and that the asymmetric perturbation from the steroidal skeleton is negligible.

Table 3-3.

			$\Delta\varepsilon_1/\Delta\varepsilon_2$
			= -2.0
		320 (-37.6)	
	309 (53,300)	307 (0.0)	CD (1),(2)
		295 (+19.2)	
	228 (12,100)		X ray (3)
4	EtOH	10% dioxane/ EtOH	

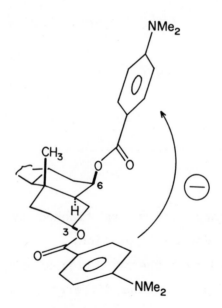

Figure 3-3. Negative exciton chirality between two p-dimethylaminobenzoate groups in the system of 5α-cholestane-3β,6β-diol bis(p-dimethylaminobenzoate).

In compound <u>4</u>, a 1,4-dibenzoate system, the CD spectrum exhibits split Cotton effects of negative first and positive second signs (Figure 3-4); the sign of Cotton effects agrees with the negative exciton chirality between the two benzoate chromophores (Figure 3-3).   The CD chirality method is thus applicable to not only vicinal dibenzoates but also more distant dibenzoates.

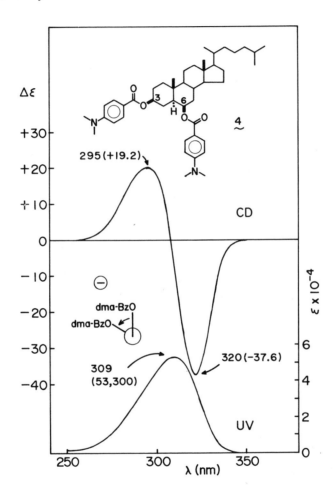

Figure 3-4. CD and UV spectra of 5α-cholestane-3β,6β-diol bis(p-dimethylamino-benzoate) in dioxane/EtOH. [Adapted from reference 1.]

Table 3-4.

			$\Delta\varepsilon_1/\Delta\varepsilon_2$
		320.6 (−36.1)	
	309.3 (58,800)	307.5 (0.0)	= −1.9
5		294.5 (+18.9)	
	226.0 (16,300)	226.0 (+1.8)	CD (5)
	EtOH	EtOH	

			$\Delta\varepsilon_1/\Delta\varepsilon_2$
		320.0 (−49.2)	= −2.7
	310.0 (60,000)	306.6 (0.0)	
6		294.5 (+18.4)	CD (5)
	226.8 (13,600)		
	0.7% dioxane/ EtOH	8% dioxane/ EtOH	

			$\Delta\varepsilon_1/\Delta\varepsilon_2$
		319 (+59.2)	= −2.0
	310 (57,500)	307 (0.0)	
7		294 (−30.2)	CD (1),(2)
	EtOH	20% dioxane/ EtOH	X ray (3)

$$\frac{\Delta\varepsilon_1}{\Delta\varepsilon_2}$$

	$310\ (55,700)$	$320\ (+28.5)$	$= -2.5$
		$307\ (\ \ 0.0)$	
		$295\ (-11.3)$	CD (1),(2)
	$227\ (14,300)$		
			X ray (3)
	EtOH	EtOH	

8

$$\frac{\Delta\varepsilon_1}{\Delta\varepsilon_2}$$

	$310\ (59,200)$	$320\ (+18.8)$	$= -2.2$
		$307\ (\ \ 0.0)$	
		$294\ (\ -8.7)$	CD (1),(2)
	$227\ (14,000)$		
			X ray (3)
	EtOH	10% dioxane/ EtOH	

9

$$\frac{\Delta\varepsilon_1}{\Delta\varepsilon_2}$$

	$310\ (54,000)$	$320\ (-35.0)$	$= -2.0$
		$308\ (\ \ 0.0)$	
		$295\ (+17.7)$	CD (1),(2)
	EtOH	EtOH	X ray (3)

10

$$\Delta\varepsilon_1/\Delta\varepsilon_2$$

319 (-20.4)   = -3.4

310 (55,800)   305 (0.0)

291 (+6.0)   CD (1),(2)

EtOH   EtOH   X ray (3)

As will be discussed later in the theoretical portion of this book, *the amplitude of split Cotton effects is inversely proportional to the square of interchromophoric distances.* Therefore, remote dibenzoates generally exhibit weaker split Cotton effects, provided the angular part remains unchanged. This tendency of decreasing $\Delta\varepsilon$ value with increasing interchromophoric distance is exemplified by the data of a series of remote dibenzoates; i.e., 1,4-dibenzoates 4-7, 1,5-dibenzoate 8, 1,6-dibenzoates 9 and 10, and 1,8-dibenzoate 11. Although compound 11 is a 1,8-dibenzoate with an interchromophoric distance of 12.8 Å, the relatively strong split Cotton effects are still observable, the sign of negative first and positive second Cotton effects being in agreement with the negative exciton chirality between the long axes of benzoate chromophores. The distance of about 13 Å is sufficiently long to encompass two remote functionalities in most organic molecules. Hence the exciton chirality method should be applicable to a wide range of compounds, provided suitable interacting chromophores are selected for derivatization.

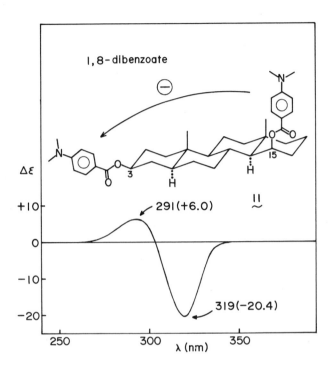

Figure 3-5. *Coupled Cotton effects of a remote dibenzoate system (1,8-gly-col); CD spectrum of D-homo-5α-androstane-3β,15β-diol bis(p-dimethylaminoben-zoate) in EtOH [adapted from reference 2].*

Table 3-5.

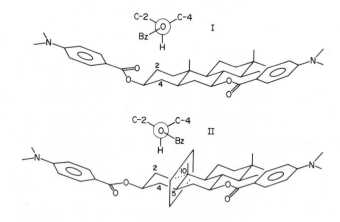

	321 (-2.8)	
310 (60,700)	*312 (0.0)*	CD (1),(2)
	300 (+4.3)	
227 (14,800)		X ray (3)
EtOH	EtOH	

12

The 3β,7β-dibenzoate *12* (Figure 3-7) is worthy of comment. The two C-O bonds are diequatorial with a dihedral angle of 0° (Figure 3-6) and hence an unsplit CD curve — i.e., summation of two independent CD curves with like or unlike signs centered at about 309 nm — might have been expected. However, although definitely much weaker than that of the 7α-epimer *8*, A_{obsd} = +39.8, the experimental A_{obsd} nevertheless had a value of -7.1. This phenomenon can

Figure 3-6. *Two possible staggered conformers of 5α-cholestane-3β,7β-diol bis (p-dimethylaminobenzoate). [Adapted from reference 2.]*

be quantitatively accounted for by assuming the 3-benzoate to adopt the two staggered conformations I and II (Figure 3-6). Conformation I leads to calculated curve I (Figure 3-7). On the other hand, when the 3-benzoate adopts conformation II, the 3- and 7-benzoates are disposed symmetrically with respect to the plane encompassing C-5/C-10/C-19 (Figure 3-6), and hence the resultant CD curve "calcd II" is nil. Summation of the two calculated curves leads to a reasonable agreement with experimental data.

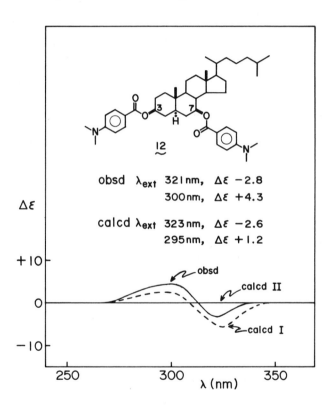

Figure 3-7. *Observed and calculated CD Cotton effects of 5α-cholestane-3β,7β-diol bis(p-dimethylaminobenzoate). Conformer I (calcd I) has a weak negative chirality, while conformer II (calcd II) has nil exciton chirality. [Adapted from reference 2.]*

Rotational Conformation of Benzoate Group. The benzoate group is capable of rotation around the alcoholic C(2)–O(3) bond,† and therefore two major conformers are conceivable for secondary alcoholic benzoates: the eclipsed and staggered conformers (Figure 3-8).   In the case of pyranose acetate, the torsional angle between the carbinyl hydrogen H(1) and ester carbon C(4) is reported to be about 30° in solution as derived from vicinal $^{13}C/^{1}H$ coupling constants.[6]   In the case of methyl formate, microwave spectroscopy showed that the staggered conformation is preferred in the gas phase; the energy barrier for internal rotation of the methyl group is 1.19 kcal/mol.[7]

In the crystalline state, the torsional angle depends on compounds; Mathieson[8] reported from X-ray data that the preferred conformation of simple esters such as acetates was eclipsed.   On the other hand, some crystalline benzoates have angles of about 30°; 25° in steroidal benzoates[9] and 30° in cocaine benzoate.[10]

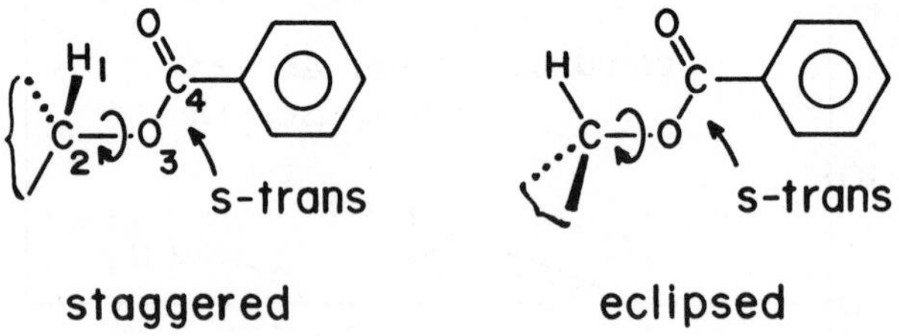

Figure 3-8. *Staggered and eclipsed conformers of secondary alcohol benzoate.*

---

† With respect to the O(3)–C(4) single bond conformation of the benzoate group (esters in general), the X-ray data of a number of compounds indicate that the phenyl (or alkyl) group is transoid and not cisoid to the C(2)–O(3) bond.

From these results it seems reasonable to consider that the torsional angle of benzoate group in solution is about 30°.   In general, however, the theoretically calculated CD values are only slightly affected by changes in torsional angles.[11]

3-1-B. Para-Substituent Effect of CD Cotton Effects of Steroidal Dibenzoates[11]

The wavelength and amplitudes of Cotton effects due to chiral exciton coupling between two benzoate chromophores depend on the substituent in the para-position. As discussed in Chapter 2, the intramolecular charge transfer transition shows a red shift as the electron-donating or -withdrawing character of the substituent increases. Consistent with this red shift in UV maxima, CD Cotton effects also exhibit red shifts, as exemplified by a series of various para-substituted benzoates of 5α-cholestane-3β,6β-diol listed in Table 3-6. The λ_{max} values of para-substituted benzoates cover the range of 230–310 nm. Therefore, if some chromophore disturbing the observation of split CD Cotton effects is present in the original glycol, one can employ a suitably para-substituted benzoate to avoid overlap of split Cotton effects with the Cotton effect of the chromophore. Actual examples will be discussed later.

Para-substituents affect not only the wavelength but also the amplitude of split Cotton effects. As mentioned in Chapter 1, chiral exciton coupling between two strongly absorbing chromophores generates strong split CD Cotton effects. A clear corroboration of this theoretical expectation is shown in Figure 3-9. Namely, the A-value, characteristic of split type Cotton effects

<u>Table 3-6.</u>

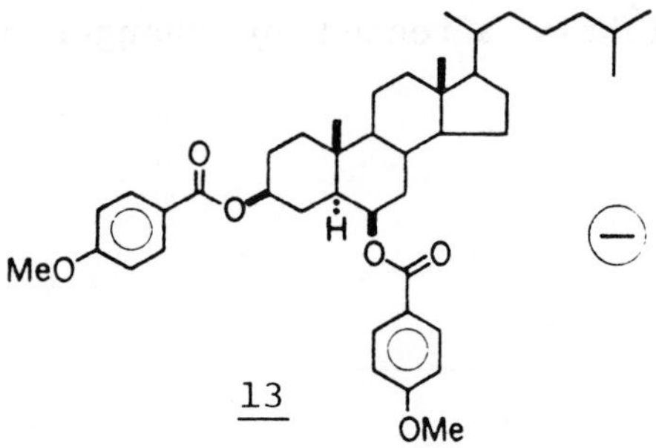

	265 (−15.1)	
	254 (0.0)	CD (11)
	243 (+11.3)	
		X ray (3)
	10% dioxane/ EtOH	

<u>13</u>

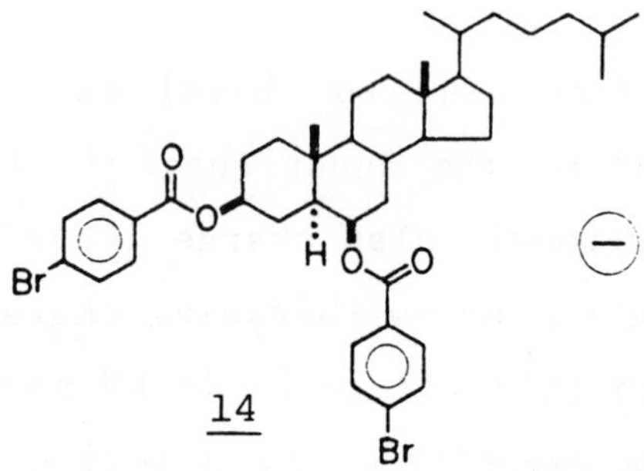

244.5 (38,000)

0.3% dioxane/
EtOH

250 (−22.5)
241 (0.0)   CD (5)
235 (+11.7)

                X ray (3)

10% dioxane/
EtOH

<u>14</u>

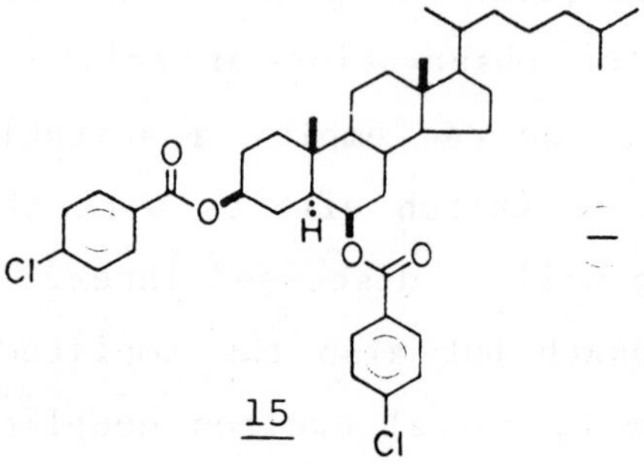

246 (−16.4)
237 (0.0)   CD (4),(11)
231 (+12.1)

                X ray (3)

10% dioxane/
EtOH

<u>15</u>

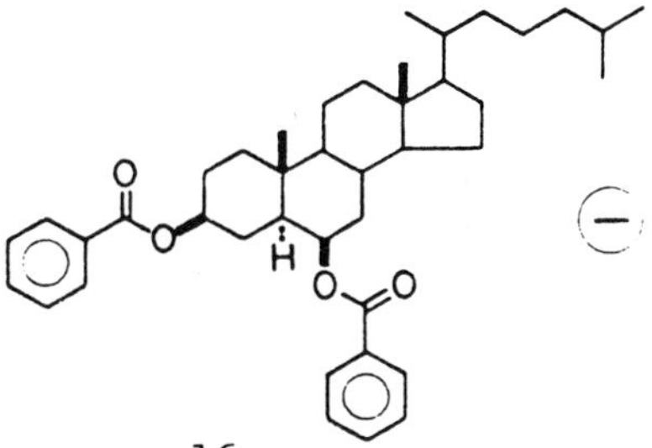

235 (−9.4)
227 (0.0)   CD (11)
220 (+6.1)
                  X ray (3)

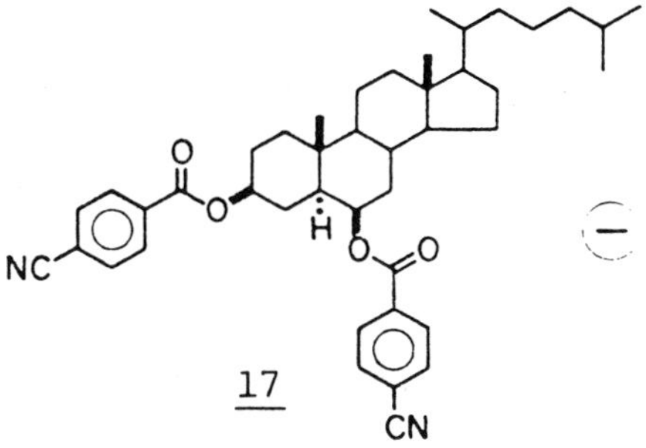

10% dioxane/
   EtOH

<u>16</u>

249 (−23.0)
237 (0.0)   CD (11)
229 (+12.4)
                  X ray (3)

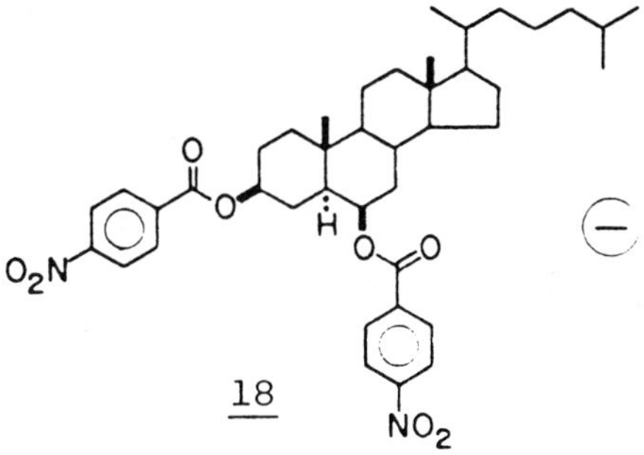

10% dioxane/
   EtOH

<u>17</u>

268 (−6.0)
256 (0.0)   CD (11)
248 (+4.7)
                  X ray (3)

10% dioxane/
   EtOH

<u>18</u>

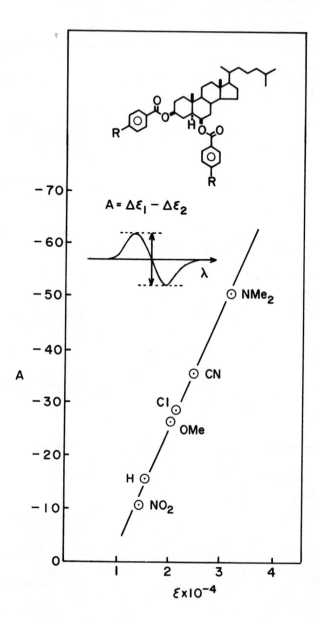

Figure 3-9. *Linear relation between A values of para-substituted dibenzoates of 5α-cholestane-3β,6β-diol and* ε_{max} *values of para-substituted benzoates of cholosterol. [Adapted from reference 11.]*

and defined as A = $\Delta\varepsilon_1$(first Cotton effect) − $\Delta\varepsilon_2$(second Cotton effect), increases as the molecular extinction coefficient value ε_{max} of the para-substituted benzoate increases.   The para-dimethylamino group generates Cotton effects of the strongest amplitude, whereas the para-nitro group gives rise to Cotton effects of weakest amplitude.   As will be discussed later, the amplitude of split Cotton effects is not only governed by the electric transition dipole moment, but is also affected by the broadness of the corresponding UV band.   For example, although the nitro group is a strong electron-withdrawing substituent, the broad UV absorption band diminishes the amplitude of split Cotton effects.

In summary, the para-dimethylaminobenzoate chromophore is the best for easy observation of split Cotton effects because of its large Cotton effect amplitude and the long wavelength location of the Cotton effect.   In general, a small amount of sample less than 1 mg is enough for CD and UV measurements.

3-1-C. Angular Dependence of Split Cotton Effects of Dibenzoates[2,12]

The exciton chirality governing the sign and amplitude of split Cotton effects is defined as follows:

$$\vec{R}_{ij} \cdot (\vec{\mu}_{i0a} \times \vec{\mu}_{j0a})\, V_{ij}$$

Since the present quadruple product is composed of vectors, exciton chirality depends on the angles between the interchromophoric distance vector $\vec{R}_{ij}$ and two electric transition moments $\vec{\mu}_{i0a}$ and $\vec{\mu}_{j0a}$ Therefore it is significant to determine the angular dependence of split Cotton effects.

In the case of vicinal glycol dibenzoates, the sign and amplitude are functions of dihedral angle between two benzoate chromophores. In this case, the sign of split Cotton effects remains unchanged in the range of the dihedral angle of 0° to 180°. Namely, the qualitative definition of exciton chirality discussed in Chapter 1 holds for vicinal dibenzoates having a dihedral angle of 0° to 180°.

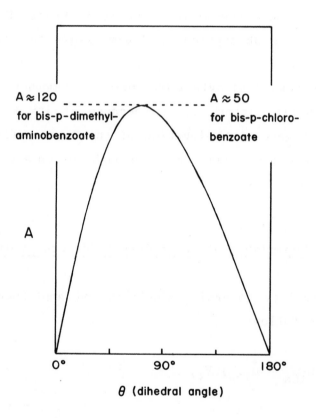

Figure 3-10. *Relation between dihedral angle* θ *and CD amplitude A* $(= \Delta\epsilon_1 - \Delta\epsilon_2)$ *of vicinal bis(p-substituted benzoates).* *[Adapted from reference 2.]*

The theoretical calculation described in Chapter 11 supports the above conclusion: *a plot of the calculated amplitude of Cotton effects of vicinal glycol dibenzoates against the dihedral angle between the two benzoate planes gives a curve having its extremum around 70°* (Figure 3-10).

This conclusion is confirmed by actual examples of vicinal dibenzoates of various dihedral angles, as shown in the following tables and figures.   The

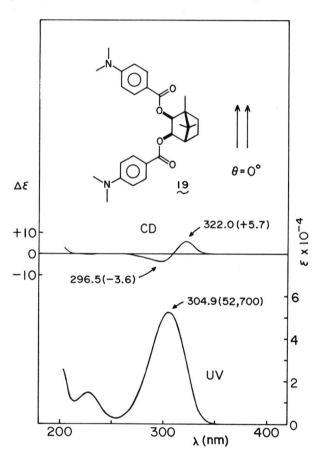

Figure 3-11.   *CD and UV spectra of (2S,3R)-camphane-2,3-diol bis(p-dimethyla-minobenzoate) in EtOH.*

Table 3-7.

		$\theta = 0°$
	322.0 (+5.7)	
304.9 (52,700)	*309.0 (0.0)*	CD (5)
	296.5 (-3.6)	
227.6 (14,900)	228.0 (-0.5)	X ray
		(14),(15)
EtOH	EtOH	

19

dihedral angles of bis(p-substituted benzoates) 19–28, as estimated from molecular models, vary from θ= 0° to 180° as indicated in the tables.   All signs
of split Cotton effects are in agreement with those predicted from the exciton
chirality method.

Table 3-8.

	$\theta \simeq +15°$
	estimated
247 (+14.9)	from
239 (0.0)	molecular
232 (-8.2)	model.
EtOH	CD (12)
	X ray (3)

20

In the case of compound <u>20</u>, the split Cotton effects are mainly due to the chiral exciton coupling between the 16α- and 17α-benzoate chromophores, the contribution of the 3-benzoate group being negligible because of the remote distance between the 3-position and 16α- or 17α-positions.

<u>Table 3-9.</u>

		θ ≃ -60°
	247 (-18.9)	
239.5 (36,000)	238 (0.0)	CD (4)
	229 (+21.0)	
		X ray (3)
10% dioxane/ MeOH	10% dioxane/ MeOH	

		θ ≃ -105°
	246 (-21.9)	
240 (37,500)	238 (0.0)	CD (12)
	231 (+21.1)	
		X ray (3)
EtOH	EtOH	

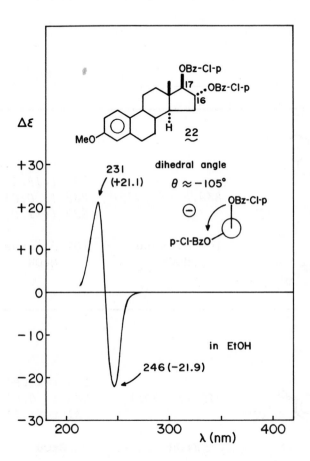

Figure 3-12. *CD spectrum of estriol 3-methyl ether 16α,17β-bis(p-chroroben-zoate) in EtOH.*

Table 3-10.

23 R = benzoyl-Cl	⊕	*248 (+36)* *230 (−11)* EtOH	θ ≃ +120° CD (13) X ray (14),(15)

24 R = benzoyl-Cl	⊖	*248 (−35)* *230 (+9)* EtOH	θ ≃ −120° CD (13) X ray (14),(15)

25	⊕	282.6 (1,800) 275.0 (2,200) 232.6 (29,500) 1% dioxane/ hexane	284.0 (−0.3) 277.2 (−0.2) 240.0 (+27.5) 228.3 (0.0) 223.2 (−5.7) 10% dioxane/ hexane	θ ≃ +120° CD (4) X ray (16)

261 (+14.0)

245 (-4.4)

EtOH

θ ≃ +120°

CD (5)

X ray (3)

The CD amplitude of compound <u>26</u>, a 1,5-dibenzoate, is smaller than those of 1,2- dibenzoates with a dihedral angle of 120° because of the remote distance between the two chromophores.

Table 3-11.

280.0 (1,500)
272.5 (1,900)

230.0 (27,600)

EtOH

236.5 (+8.8)
224.5 (0.0)
221.0 (-0.9)

EtOH

θ ≃ +180°

CD (4)

X ray (3)

		$\theta \simeq +180°$
	321.0 (+12.2)	
314.8 (61,500)	308.5 (0.0)	CD (5)
	297.0 (-5.8)	
228.0 (14,400)	232.0 (-0.9)	X ray (3)
EtOH	EtOH	

Compounds <u>27</u> and <u>28</u> are also worthy of comment: the 2β- and 3α-benzoates are diaxial in the ideal chair conformation for the steroidal A-ring. Since both benzoate groups make a dihedral angle of 180°, no split type Cotton effects should result. However, compound <u>27</u> exhibits a positive first Cotton effect while the second Cotton effect is not observed because it is overlaid by the background ellipticity. The present data are assignable as follows: the 1,3-steric hindrance exerted by the 19-methyl group distorts the 2β-benzoate towards the outside of the A-ring leading to a positive chirality between the two benzoate groups, which is in accordance with the observed Cotton effect sign. In fact, the CD spectrum of bis(<u>p</u>-dimethylaminobenzoate) <u>28</u> corroborates the assignment: it clearly exhibits relatively weak but typical split type Cotton effects of positive first and negative second signs, in agreement with the above assignment (Figure 3-13).

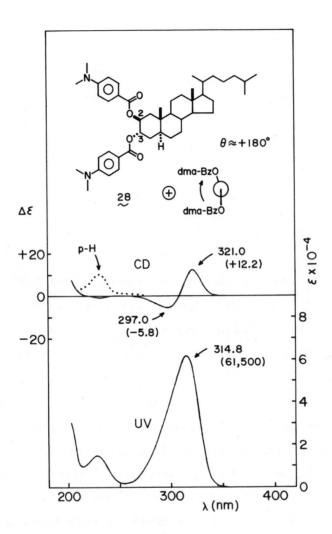

Figure 3-13. CD and UV spectra of 5α-cholestane-2β,3α-diol   bis(p-dimethylaminobenzoate) in ethanol. The dotted line shows the CD curve of the corresponding unsubstituted dibenzoate in ethanol.

As exemplified in various dibenzoates discussed above, the visualized definition of exciton chirality is convenient and simple for assignments of split Cotton effects.   However, it should be remembered that in some special cases such as 1,1'-binaphthyl[17,18] 9,9'-bianthryl, etc., theory predicts that the sign of split Cotton effects changes with the dihedral angle within the range of 0–180°; the visualized definition of exciton chirality is not applicable.   In such cases, it is important to examine the quantitatively defined exciton chirality; i.e., the quadruple product, $\vec{R}_{ij} \cdot (\vec{\mu}_{i0a} \times \vec{\mu}_{j0a}) V_{ij}$.   Although this is seldom required, it is obvious that assignments based on quantitative definition of exciton chirality are superior to those based on the qualitative definition.

The CD and UV data of some other steroidal dibenzoates are listed in Table 3-12.

Table 3-12.

	272.4 (1,900)	280.5 (+0.7)	
		273.5 (+0.8)	
		238.5 (−15.2)	
	228.2 (26,700)	230.7 (0.0)	CD (4)
		223.0 (+15.7)	
29			X ray (3)
	EtOH	EtOH	

	280.0 (1,400)	281.0 (-0.4)	
	272.8 (1,800)	274.0 (-0.3)	
		237.5 (+19.6)	
(+)	*228.2 (26,200)*	227.5 (0.0)	CD (4)
		222.0 (-8.5)	
			X ray (3)
	EtOH	EtOH	

_30_

		251 (-28.7)	
(−)	*244.5 (39,900)*	243 (0.0)	CD (5)
		235 (+20.6)	
			X ray (3)
	0.3% dioxane/	10% dioxane/	
	EtOH	EtOH	

_31_

	280.2 (1,500)		
	272.8 (1,800)		
		255.0 (-1.3)	
		237.8 (+20.1)	
(+)	*229.0 (27,000)*	229.0 (0.0)	CD (5)
		222.8 (-11.6)	
			X ray (3)
	0.9% dioxane/	10% dioxane/	
	EtOH	EtOH	

_32_

279.8 (1,400)		
272.3 (1,800)		
	237.0 (+17.5)	
228.7 (26,900)	227.2 (0.0)	CD (5)
	221.2 (-5.6)	
		X ray (3)
0.9% dioxane/	10% dioxane/	
EtOH	EtOH	

33

___

3-1-D. Dependence of Exciton Cotton Effects on the λ_{max} Separation of Two Different Chromophores[5]

Exciton coupling Cotton effects are characteristic of binary systems with two identical chromophores, as discussed in the previous sections. Therefore, when applying the exciton chirality method, it is advantageous to employ two identical chromophores. However, the exciton method is still applicable to nondegenerate systems composed of two different chromophores having similar λ_{max} positions. Figure 3-14 shows theoretical results of the dependence of exciton Cotton effects on the λ_{max} separation (see Chapter 10 for the theoretical calculation). It indicates that the CD amplitude gradually decreases as the λ_{max} separation of two chromophores increases and that the CD exciton chirality method is applicable to chromophores with different λ maxima.

The experimental data shown in Figure 3-15 support these theoretical results. Namely, the unsubstituted benzoate and p-methoxybenzoate system still exhibits exciton split Cotton effects of moderate intensity. The

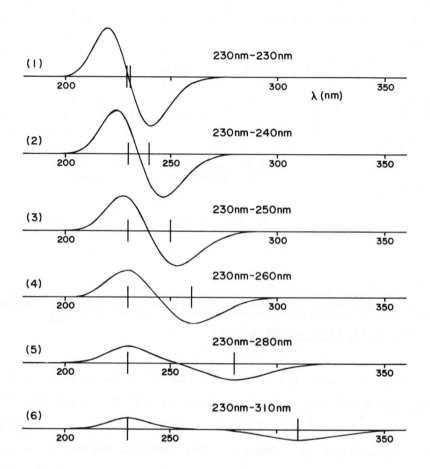

Figure 3-14. *Theoretical results on the dependence of exciton Cotton effects on* λ_{max} *separation of two different chromophores. CD curve (1) is the case of two identical chromophores. Curve (2) is of a non-degenerate system in which 230 nm and 240 nm transitions interact with each other. In the calculation, only the value of* λ_{max} *separation was varied while remaining parameters were kept constant. The following values were adopted:* $V_{ij} = 240 \ cm^{-1}$ *and* $\Delta\sigma = 2700 \ cm^{-1}$.

case of the benzoate/p-dimethylaminobenzoate system is also worthy of comment;
the negative Cotton effect at the long wavelength of about 310 nm is completely separated from the shorter wavelength Cotton effects and appears to be an
independent extremum.   However, comparisons of the CD curves depicted in Figures 3-14 and 3-15 show that the negative sign of the Cotton effect is still
governed by the negative exciton chirality of the coupled system.

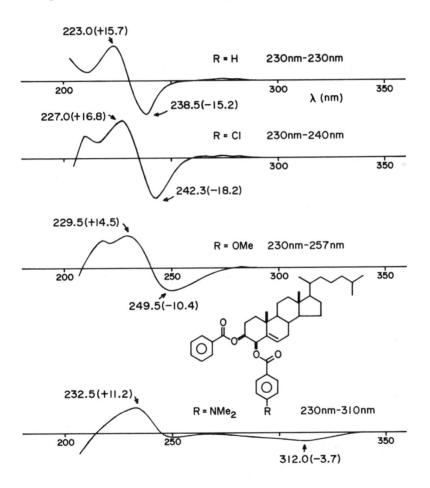

Figure 3-15. *The observed CD spectra of cholest-5-ene-3β,4β-diol   3-benzoate*
4-p-substituted benzoates in ethanol.[5]

3-1-E.   Angular Dependence of UV λ_{max} of Exciton Coupling Systems[80]

As discussed in previous sections, when two identical chromophores couple with each other, the CD spectrum exhibits split Cotton effects of opposite signs that are characteristic of exciton coupling systems. On the other hand, the UV spectrum shows no remarkable change; the UV curve merely looks like an unsplit absorption band of double intensity (for example, compare Figure 1-6 with Figure 1-5). However, careful examination of UV spectra shows that UV λ_{max} of exciton coupling systems depends on the angle between the two dipole moments; *when the angle is less than 90°, the UV spectrum exhibits a blue shift*, in comparison with the λ_{max} of monomer. On the other hand, *when the angle is larger than 90°, the UV spectrum shows a red shift.*

For example, in the case of the p-dimethylaminobenzoate chromophore, the λ_{max} of monobenzoate is 311.0 nm. The UV spectrum of a binary system, (2S,3R)-camphane-2,3-diol bis(p-dimethylaminobenzoate) 19, in which the dihedral angle between the two benzoate groups is 0°, exhibits a blue shift of 6.1 nm (λ_{max} 304.9 nm, Figure 3-11). On the other hand, the UV spectrum of 5α-cholestane-2β,3α-diol bis(p-dimethylaminobenzoate) 28 with a dihedral angle of about 180° shows a red shift of 3.8 nm (λ_{max} 314.8 nm, Figure 3-13).

The present results can be interpreted on the basis of the exciton coupling mechanism, as follows: if the angle between two transition moments is about 0°, the out-of-phase combination of the two transition moments (α-state) is energetically more stable than the in-phase combination (β-state), as illustrated in the upper part of Figure 3-16. The in-phase combination at shorter wavelengths gives rise to a more intense absorption band because the UV intensity is proportional to the square of sum of vectors. On the other hand, the out-of-phase combination at longer wavelengths gives rise to a weak absorption band.   Therefore, summation of the two curves results in a blue shift of UV λ_{max}.   If the angle is approximately 180°, the order of energy levels is reversed; the β-state of in-phase combination is stabler than the α-state of out-of-phase combination, as shown in the lower part of Figure 3-16.   Therefore, an intense absorption band is located at longer wavelengths while a weak band is at shorter wavelengths;   summation of the two curves

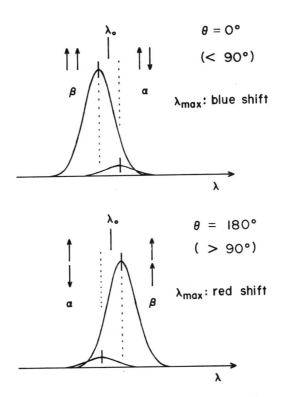

Figure 3-16.   *Angular dependence of UV λ_{max} of exciton coupling systems: when the angle between two dipole moments is less than 90°, the UV spectrum exhibits a blue shift relative to the monomer.   On the other hand, an angle larger than 90° leads to a red shift. [Adapted from reference 80.]*

Table 3-13. Angular Dependence of UV λ_{max} of p-Dimethylaminobenzoates.

Compound	λ_{max} (nm)	Solvent[a]
Monobenzoate		
	311.0	D/E
Dibenzoate		
0° (1,3-)	302.6	EtOH
40° (1,2-)	304.9	EtOH
70° (1,2-)	307, 308, 308, 309.4	EtOH
70° (1,4-)	309, 309.3, 310.0, 310	EtOH
70° (1,5-)	310, 310	EtOH
90° (1,3-)	310.5, 311.8	EtOH
120° (1,3-)	312.0	EtOH
180° (1,3-)	313.9	D/E
180° (1,2-)	314.8	EtOH

[a] D/E: 1% dioxane in EtOH

results in a red shift. The same discussion also holds on the basis of the equations formulated in Table 10-2.

The UV data of other pertinent bis(p-dimethylaminobenzoates) with various dihedral angles and distances are listed in Table 3-13; as the dihedral angle increases, UV λ_{max} gradually shifts to longer wavelengths. Provided the angle remains unchanged, a dibenzoate system with longer interchromophoric distance exhibits a smaller shift value because of weak exciton interaction. All of the data listed in the table are in agreement with the present general rule based on the exciton theory. From UV λ_{max} data, in general, one can discriminate whether the angle between two transition moments is less than 90° or larger than 90°, while CD data enable one to determine the absolute screwness between two transition moments.

3-2. Summary of Practical Aspects and Requirements to be Considered in Applying the Exciton Chirality Method

1. The interaction between two p-dimethylaminobenzoates is still strong at a distance of 12.8 Å , as exemplified by the interaction between two groups located at 3β and 15β positions (1,8-dibenzoate) in a D-homosteroid (Figure 3-5). This distance is long enough to encompass most natural products.

2. In the case of vicinal dibenzoates, the signs of split Cotton effects remain unchanged in the range of dihedral angle of 0° to 180°. It is pertinent to ascertain relative configuration and/or conformation by other physical methods, such as NMR J-values.

3. In principle, any chromophore possessing intense $\pi \rightarrow \pi *$ electronic bands can be used as components of interacting chromophores which give rise to split CD curves. The following are frequently encountered in various natural products themselves:

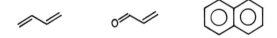

Table 3-14.   UV λ_{max} Position of Pertinent Para-substituted
Monobenzoates and CD λ_{ext} Positions of the Corresponding Dibenzoates (in EtOH).

X	λ_{max} (nm)	1st Cotton (nm)	2nd Cotton (nm)
NMe$_2$	311.0	320	295
OMe	257.0	265	243
Br	244.5	250	235
Cl	240.0	246	231
H	229.5	235	220
CN	240.0	249	229
NO$_2$	260.5	268	248

Provided the λ_{max} of these chromophores are not too far apart,
their interaction results in exciton split CD curves (see Chapters 4,
6, and 8).   If the sample contains only one such chromophore but an
additional hydroxyl group, then the latter is converted into para-
substituted benzoates having their λ_{max} at a wavelength close to
the existing chromophore (see Table 3-14).

4. In compounds which contain two or more hydroxyls, the benzoate group should be chosen so that the λ_{max} is as far removed as possible from the λ_{max} of the already present chromophore (if any). This gives rise to less ambiguous results because we are now dealing with interaction between two or more identical chromophores.

5. The benzoate groups that have been used for practical purposes are listed in Table 3-14 together with their λ_{max} and the positions of the two split CD extrema resulting from interaction between two benzoate groups.

Table 3-15.

280.2 (1,400)		
272.9 (1,700)		
	236.0 (-19.0)	
229.0 (23,900)		CD (5)
0.9% dioxane/ EtOH	10% dioxane/ EtOH	X ray (3)

6. In some cases, the second Cotton effect which is usually weaker than the first Cotton effect is masked by a strong background ellipticity. For example, cholest-5-ene-3α,4α-diol dibenzoate (**34**) exhibits only one CD extremum at 236 nm as the first Cotton effect, while the second Cotton effect is buried in a strong negative background ellipticity (Figure 3-17). In such cases, the judgment as to whether an observed Cotton effect belongs to the exciton split type or not is

based on the position of the first Cotton effect of the particular benzoate group. If we are dealing with an unsubstituted dibenzoate, the observation of a Cotton effect around 235 nm indicates that it belongs to this class; in contrast, if it is around 230 nm, it is estimated that the Cotton effect is not due to exciton interactions.

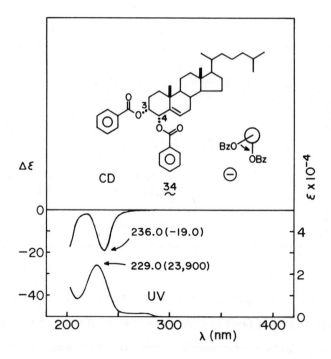

Figure 3-17.   *CD and UV spectra of cholest-5-ene-3α,4α-diol dibenzoate in dioxane/EtOH.*

7. The $\Delta\varepsilon$ values of the two split Cotton effects are not necessarily equal, that of first Cotton effect being more intense. However, the integrated peak areas of the two coupled bands are approximately equal to each other. In this case, the so-called "Sum Rule" holds for the region of exciton coupling (see Figure 3-1 and section 11-3).

8. In a few cases,[17-19] depending on the geometry of molecules, a visualized definition such as the Newman projection does not lead to a clearcut prediction of chirality; in such cases, the quadruple product, $\vec{R}_{ij} \cdot (\vec{\mu}_{i0a} \times \vec{\mu}_{j0a}) V_{ij}$, of the quantitative definition should be examined.

3-2-A. Requirements of the Exciton Chirality Method

In the CD exciton chirality method, it is important to choose the proper electronic transition of proper chromophores satisfying the following requirements of chiral exciton coupling:

1. *large extinction coefficient values in UV spectra;*
2. *isolation of the band in question from other strong absorptions;*
3. *established direction of the electric transition moment in the geometry of the chromophore;*
4. *unambiguous determination of the exciton chirality in space, inclusive of configuration and conformation;*

5. *negligible molecular orbital overlap or homoconjugation between the chromophores, if any.*

3-2-B. Recommendation for Using the p-Dimethylaminobenzoate Chromophore

The para-dimethylaminobenzoate chromophore is the best for easy observation of split Cotton effects and reliable assignment for the following reasons:

1. The CD Cotton effects are located around 310nm, a long wavelength region usually devoid of interfering absorptions.

2. The $\Delta\varepsilon$ values are very large ($\Delta\varepsilon$ 70 ∿20). Therefore, the exciton split Cotton effects are clearly observable without overlap with other Cotton effects. For example, cholest-5-ene-3α,4α-diol bis(p-dimethylaminobenzoate) (35) exhibits strong exciton split

Table 3-16.

		$\Delta\varepsilon_1/\Delta\varepsilon_2$
	321.2 (-67.6)	
309.4 (55,000)	307.2 (0.0)	= -2.7
	295.5 (+24.6)	
227.8 (14,900)	234.5 (-2.6)	CD (5)
EtOH	EtOH	

35

Cotton effects unlike the case of the unsubstituted dibenzoate
(compare Figures 3-18 and 3-17). In the latter case, the second
Cotton effect is masked by a strong background ellipticity.

3. Preparation of p-dimethylaminobenzoate is relatively easy.

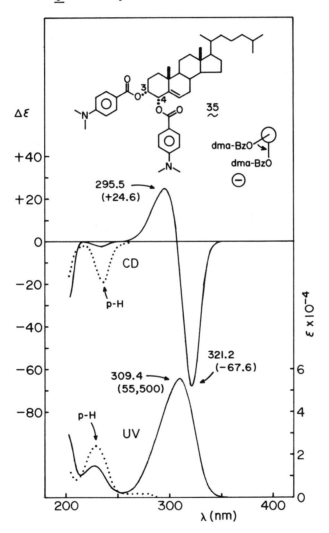

Figure 3-18. *CD and UV spectra of cholest-5-ene-3α,4α-diol   bis(p-dimethyla-*
minobenzoate) in EtOH.

3-2-C.   Preparation of p-Dimethylaminobenzoates

In general, p-dimethylaminobenzoates can be prepared by heating alcohols with p-dimethylaminobenzoyl chloride in pyridine.   p-Dimethylaminobenzoyl chloride is prepared as follows, according to the literature procedure.[20,21] A solution of oxalyl chloride (1.6 g, 12.6 mmol) in dry benzene (3 mL) is added slowly to a mixture of potassium p-dimethylaminobenzoate (2.5 g, 12.3 mmol) and dry benzene (10 mL) at 0°C, stirred at room temperature for 20 minutes, and refluxed for 1 hour.   After cooling, removal of potassium chloride by filtration, and concentration precipitate p-dimethylaminobenzoyl chloride, mp 146–148 °C: lit.,[20] mp 147–148 °C.

An alternative, more convenient, preparation is as follows.[21]   A suspension of p-dimethylaminobenzoic acid (10 g, 60 mmol) in CS_2 (100 mL) containing pyridine (5.11 mL, 63 mmol) is treated slowly with a suspension of PCl_5 (12.5 g, 60 mmol) in CS_2 (200 mL) at room temperature.   The reaction mixture is refluxed with stirring until the white solid has dissolved completely.   The remaining yellow orange precipitate is removed by filtration and the hot filtrate is allowed to cool to room temperature, giving p-dimethylaminobenzoyl chloride as white plates (8.0 g, 72%, mp 145–147 °C: lit.,[21] mp 147–148 °C).

3-3.   Application of the Dibenzoate Chirality Method for Determining Absolute Stereochemistries of Glycols

3-3-A.   Natural Products

The dibenzoate chirality method has been successfully applied to various natural products to determine the absolute configurations and/or conformations, as exemplified in the following.

Table 3-17.

	327 (+1.8)	
	248 (−3.9)	
	235 (−14.5)	
231 (32,300)	227 (0.0)	CD (4)
	218 (+15.9)	
		X ray
EtOH	EtOH	(3),(22)

36

In the case of dibenzoate **36** of ponasterone A , a phytoecdysone, the Cotton effects of the enone group (n→π*, λ_{ext} 327 nm, Δε= +1.8; π→π*, λ_{ext} 248 nm, Δε= −3.9) are weaker than those of the dibenzoate, and hence the Cotton effects due to the 2β,3β-dibenzoate moiety are dominating (λ_{ext} 235 nm, Δε= −14.5 and λ_{ext} 218 nm, Δε= +15.9).   In comparison with the strong 1,2-dibenzoate interaction, the enone-benzoate interaction should be negligible because of the remote distance and different excitation wavelengths.   Based on these data, it is concluded that compound **36** has its A-ring (A/B cis) in the chair conformation.[5]

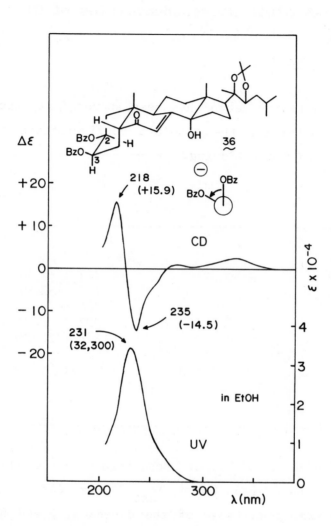

Figure 3-19. CD and UV spectra of dibenzoate of ponasterone A derivative in EtOH.

Table 3-18.

236 (-16.4)	CD (23)
	X ray
221 (+12.1)	(3),(22)
EtOH	

In a similar way, the absolute configuration of the $2\alpha,3\alpha$-glycol   moiety in ponasterone B (dibenzoate <u>37</u>) has been determined.[23]

Table 3-19.

236 (-14.6)	CD (23)
220 (+14.4)	X ray
	(3),(22)
EtOH	

The present method was also applied to establish the $2\beta,3\beta$-glycol   moiety in ponasterone C.[23]   In this case, since the tetra-benzoate derivative <u>38</u> was employed for CD measurement, the observed Cotton effects are affected by the benzoate groups in the side chain.   However, it is safe to conclude that the split Cotton effect is dominated by the ring A 1,2-dibenzoate moiety and not by the acyclic 1,4-dibenzoate because of its longer distance and conformational mobility.

The absolute configuration of the glycol moiety in rishitin (dibenzoate **39**), an antifungal norsesquiterpene, was determined as depicted in Table 3-20 and Figure 3-20 on the basis of this method.[4] The absolute stereochemistry was confirmed by the synthesis from (-)-α-santonin.[24]

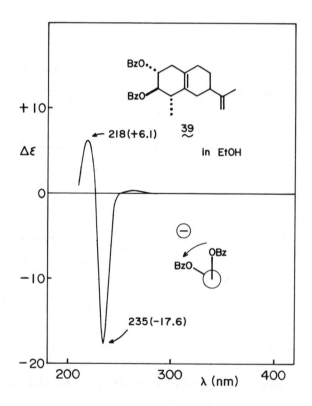

Figure 3-20. CD spectrum of rishitin dibenzoate in EtOH.

Table 3-20.

		235 (−17.6)	
229.8 (24,300)		226 (0.0)	CD (4)
		218 (+6.1)	
			Correl. (24)
EtOH		EtOH	

39

Table 3-21.

235 (−19.2)	CD (25)
219 (+5.9)	X ray, relative
EtOH	(26)

40

Masamune and coworkers[25] determined the absolute configuration of oxylubimin (dibenzoate **40**), a sesquiterpene from diseased potato tubers, in a similar manner. The result is in accord with that expected from the biosynthetic correlation with rishitin.

In the case of tetrahydrotaxinine 9,10-dibenzoate (<u>41</u>), the highly strained enone group shows an usually strong $\pi{\rightarrow}\pi^*$ Cotton effect at 266 nm and a weaker $n{\rightarrow}\pi^*$ Cotton effect at 350 nm, but the dibenzoate Cotton effects are clearly measurable (asterisked peaks in Figure 3-21).[4]

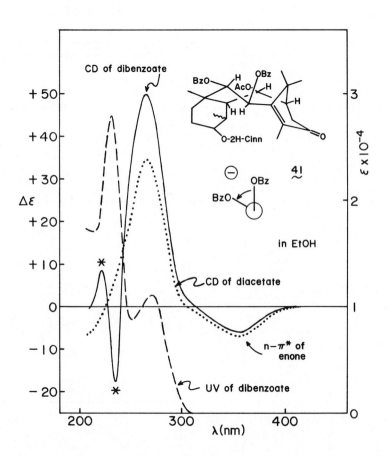

Figure 3-21. *CD and UV spectra of dibenzoate of taxinine derivative in EtOH. The dibenzoate Cotton effects marked by *:   ——— CD of dibenzoate; ⋯⋯⋯ CD of diacetate;   ----- UV of dibenzoate. [Adapted from reference 4.]*

Table 3-22.

		350 (−5.7)	
		266 (+49.8)	
	270.0 (11,100)	*234 (−17.6)*	
		227 (0.0)	CD (4)
	230.8 (27,700)	222 (+8.5)	
			X ray
	EtOH	EtOH	(27),(28)

41

Table 3-23.

		270 (−29.9)	
		258 (0.0)	CD (29)
	250.0 (44,100)	*247 (+30.8)*	
		223 (−20.0)	Correl. (30)
	EtOH	EtOH	

42

As mentioned earlier, if the glycol itself contains some chromophore exhibiting relatively strong Cotton effects in the region of the nonsubstituted benzoate transition, one can prevent overlap of benzoate Cotton effects with the original Cotton effect by employing a para-substituted benzoate chromophore. For example, dimethylbergenin exhibits a negative Cotton effect arising from the gallate chromophore at 224 nm (Figure 3-22), which would overlap with the split Cotton effects of unsubstituted benzoates ($\Delta\epsilon$ 10 ∿ 15 at

235 nm and 220 nm).   On the other hand, extrema of the bis(p̲-methoxybenzoate)
42 are not only located at 270 and 247 nm, where there would be no overlap,
but also have very strong amplitudes and thus can be clearly distinguished
from the twist-gallate Cotton effect at 224 nm (Figure 3-22).[12,29]   In order
to obtain clearer results, however, a p̲-dimethylaminobenzoate chromophore
would have been preferred.

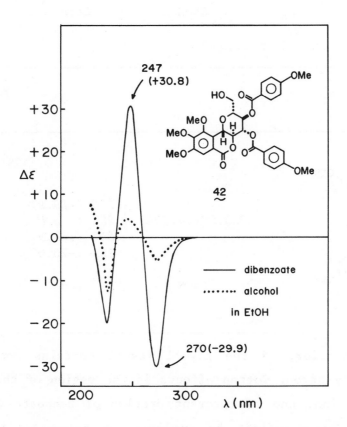

*Figure 3-22. CD spectra of dimethylbergenin and its bis(p̲-methoxybenzoate) in
EtOH (the phenolic group para to the lactone is the one methylated in the
original bergenin). [Adapted from reference 12.]*

Table 3-24.

		248 (−18.2)	
242.5 (35,500)		239 (0.0)	CD (31)
		229 (+12.6)	
EtOH		EtOH	X ray (32)

43

The absolute configuration of illudin S (44), an anti-tumor antibiotic, had remained to be established in spite of several attempts to elucidate it by chemical or spectroscopic methods.   In order to apply the dibenzoate chirality method, illudin S 44 was converted into a phenol derivative 46, and this was acylated with p-chlorobenzoate chromophores.   The CD spectrum of bis (p-chlorobenzoate) 43 shows typical negative first and positive second Cotton effects, which indicates that a negative exciton chirality should be assigned to the two benzoate chromophores in compound 43, as illustrated in Figure 3-23.[31]. The positions of the two split Cotton effects are typical for

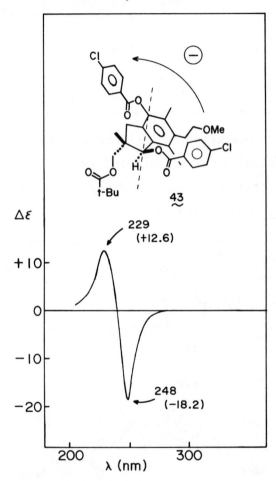

Figure 3-23. CD spectrum of bis(p-chlorobenzoate) of illudin S derivative in EtOH. [Adapted from reference 31.]

bis(p̲-chlorobenzoates) and this demonstrates that the effect of the phenolic moiety is negligible because of the nonplanarity between this chromophore and the attached p̲-chlorobenzoate chromophore (steric hindrance due to ortho methyl and peri-methylene).   The absolute configuration of illudin S thus established was subsequently confirmed by the X-ray Bijvoet method.

Table 3-25.

47 ⊕	233 (33,900)	241 (+16)	CD (33)
		224 (-19)	
	MeOH	MeOH	

In a similar way, Natori and coworkers[33] have deduced the absolute configuration of shinanolone by measuring the CD Cotton effects of dibenzoate **47**.   In order to arrive at an unambiguous assignment, they attempted to remove the carbonyl group which contributes as a twisted tetralone chromophore and disturbs the observation of genuine split Cotton effects due to benzoates.   A more clearcut split Cotton effect, without removal of the carbonyl group, could possibly have been observed by employment of para̲-substituted benzoate chromophores which shift the Cotton effects to wavelengths beyond the tetralone absorbing region.

Table 3-26.

	324 (11,000) 241 (28,800) EtOH	*250 (+59.2)* *241 (0.0)* *230 (−74.2)* EtOH	CD (34) X ray (35)
	324 (11,000) 245 (30,900) EtOH	*249 (−56.5)* *240 (0.0)* *232 (+51.3)* EtOH	CD (34) X ray (35)

On the basis of NMR and CD studies, Yamada, et al.[34] reported the conformations of (−)-cis-khellactone and (+)-trans-khellactone of known absolute configurations. Bis(p-chlorobenzoate) 48 exhibits typical strong positive first and negative second Cotton effects due to the dibenzoate moiety, which indicates positive chirality. The present exciton coupling between the two benzoate groups is not affected by the transition at 324 nm of the coumarin chromophore, because the transition is located far from the benzoate transition. In the case of compound 49, the exciton chirality between two benzoate groups is always negative, irrespective of the conformation of the ring.

The absolute configuration is thus unambiguously determinable from the present data, the obtained configuration being in agreement with that deduced by the X-ray Bijvoet method.

Table 3-27.

237 (-)		CD (36)
222 (+)		Correl. (36)
		X ray (37)

50

Ziffer and coworkers[36] have applied the dibenzoate chirality method for determining the absolute stereochemistry of (+)-cis-1,2-dihydroxy-3-methylcyclohexa-3,5-diene (52) produced from toluene by Pseudomonas putida. Although the dihydrodiol 52 shows a positive Cotton effect (λ_{ext} 270 nm, $\Delta\varepsilon$= +0.60), the absolute configuration could not be assigned from the CD spectrum alone by applying the diene helicity rule, because compound 52 can exist in either one of the two conformers in which the skew sense of the diene differs.

In order to apply the exciton method, the dihydrodiol 52 was hydrogenated to the saturated glycol 53, the dibenzoate 50 of which showed a negative chiroptical effect at 237 nm and a positive one at 222 nm, thus establishing the (1S,2R,3R) configuration.   This absolute stereochemistry was found to be in agreement with that obtained by chemical correlation and X-ray Bijvoet methods.

Table 3-28.

234 (-10.6)	CD (38)	
222 (+5.2)	Correl. (38)	
optical purity 40%		
isooctane		

54

Ziffer, et al.[38] also reported the application of the dibenzoate chirality method to trans-p-menthane-2,3-diol (dibenzoate 54) of lower optical purity (around 40%), and showed the utility of the exciton method for assigning the absolute configuration of monocyclic 1,2-cyclohexanediols.

Tsuda and coworkers[39] have determined the absolute configuration of the vicinal glycol moiety in lycoclavanin (dibenzoate 55), a triterpenoid tetrol of L. clavatum, by means of the dibenzoate chirality method.   Similar to the case of ponasterone A, the enone-benzoate interaction can also be neglected.

Table 3-29.

239 (+29.6)	CD (39)
dioxane/ MeOH	X ray (40)

Table 3-30.

CD (41)	
X ray (42)	

Sakakibara, et al.[41] applied the exciton method to lyoniol-D (dibenzoate 56), a toxic diterpenoid isolated from Lyonia ovalifolia Drude val elliptica Hand-Mazz, to determine the 2α,3β-absolute configuration of the vicinal glycol. However, no numerical CD data were reported.

Table 3-31.

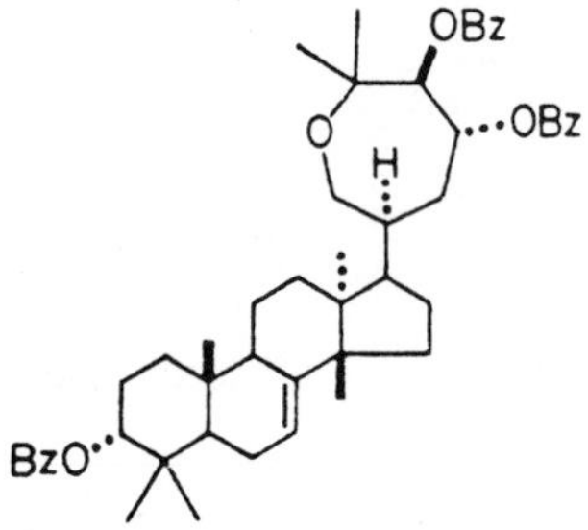

238 (-18)

CD (43)

Correl. (43)

57

Lyons and Taylor[43] have determined the absolute stereochemistry of saperin B by observing the split type Cotton effects of tribenzoate 57.   In this case, the contribution of the 3-benzoate group can be disregarded because of the long distance from the vicinal dibenzoate moiety.   The stereochemistry was also determined by chemical correlation data, but the empirical organo-metallic method employing $Pr(DPM)_3$ (see Chapter 9) led to the wrong configuration.

Table 3-32.

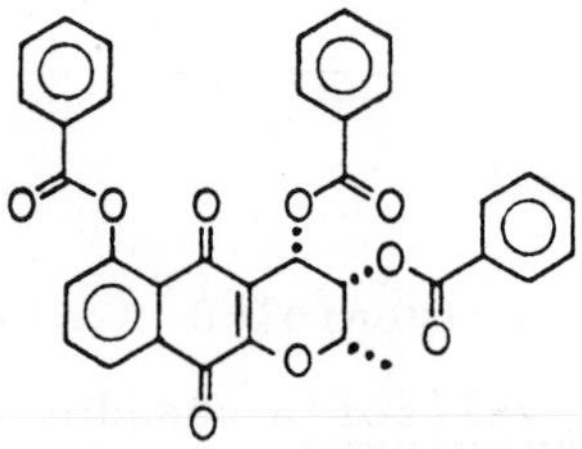

345   (3,700)
275   (16,200)

238 (+)

232.5 (50,100)

222 (-)

CD (44)

CH_2Cl_2

CH_3CN

58

Glosse, et al.[44] have elucidated the absolute configuration of cryptosporin, a metabolite of the fungus Cryptosporium pinicola Linder, by using the Cotton effects of tribenzoate 58.   However, in order to obtain a more definite conclusion, it would have been better to block the phenol group by methylation and to employ a para-substituted benzoate chromophore; this is in order to avoid interaction with the naphthoquinone chromophore.

Table 3-33.

59	315 (-8.4) 247 (+51) 232 (shoulder)	CD (45)
60	313 (+9.3) 249 (-31.7) 233 (+18.1)	CD (45)

Nukina, et al.[45] have determined the absolute configuration of radicinin (61) and (-)-radicinol (62), antibiotic metabolites isolated from Coch-

liolus lunata, by using the dibenzoate chirality method.   Reduction of radicinin 61 afforded radicinol 62 together with its epimer 63.   The cis and trans
configurations of two hydroxyl groups in 63 and 62, respectively, were
assigned on the basis of the fact that 63 easily formed an acetonide while
62 resisted.   The conformations of bis(p-chlorobenzoates) were assigned as
depicted in Figure 3-24 from ^{1}H NMR coupling constant data.   In the case of
59, the two benzoate groups are in diaxial positions, indicating the dihedral
angle of 180°.   The preferred conformation, however, results in the distortion
of the two benzoate groups towards the outside of the pyrane ring, constitut-

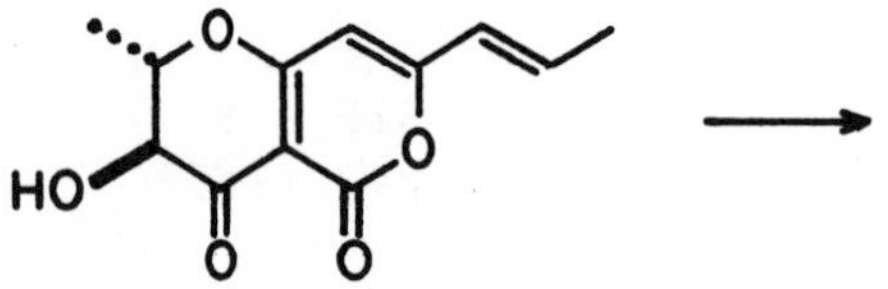

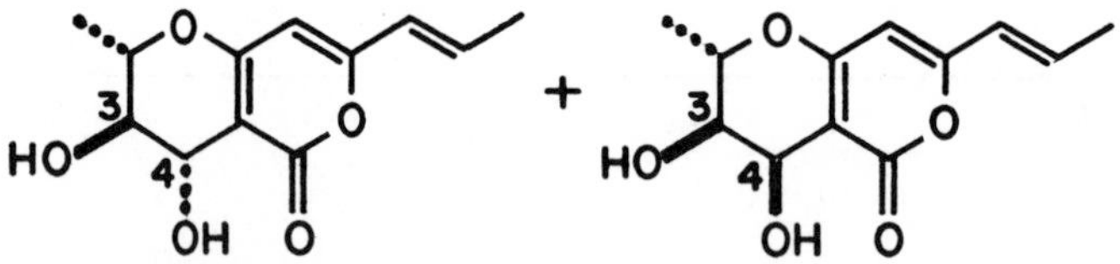

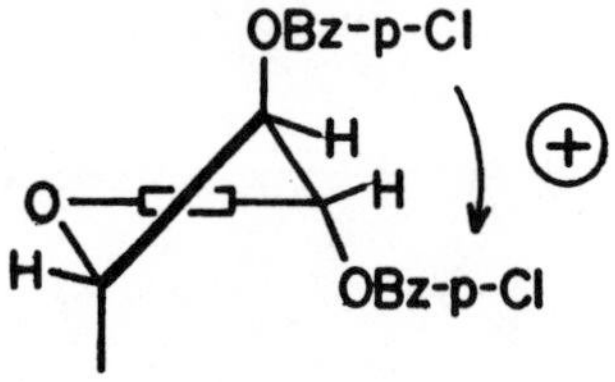

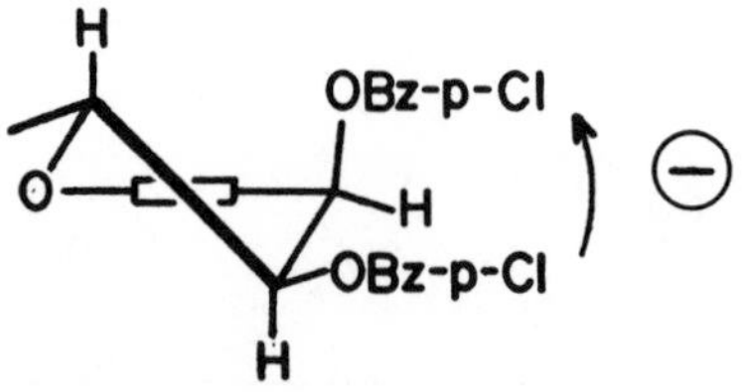

*Figure 3-24.   Conformation and exciton chirality of radicinol bis(p-chloroben-
zoate) and its epimer.*

ing a positive exciton chirality. The positive first Cotton effect clarified the (2S,3R,4S) absolute stereochemistry, although the second Cotton effect was not observed.

Table 3-34.

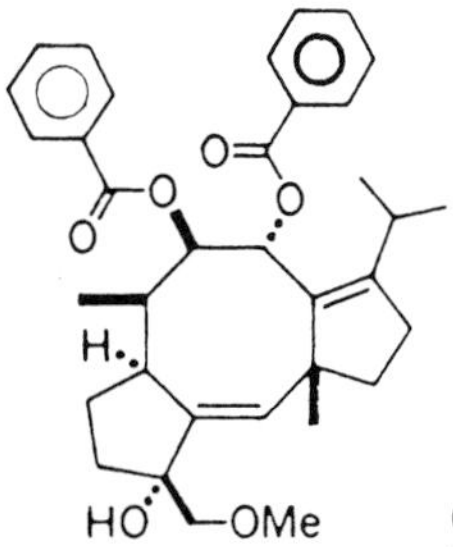

238 (-20.9)

CD (46)

64

Sassa, et al.[46] determined the absolute stereochemistry of cotylenol (dibenzoate 64), a leaf growth substance produced by a fungus, Cladosporium sp., from the exciton split CD Cotton effects of the vicinal dibenzoate system contained in an eight-membered ring.

Table 3-35.

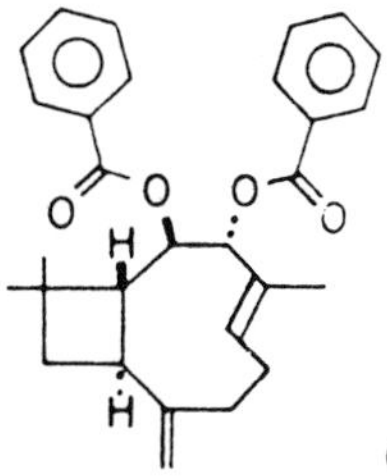

237 (-15.3)   CD (47)

223 (+19.0)   X ray
                     relative
                     (47)

65

<u>66</u> $\oplus$

236 (+20.5)

220 (-3.5)

CD (47)

X ray
relative
(47)

Similarly, Yoshida, et al.[47] determined the absolute configuration of buddledin A, a piscidal sesquiterpene isolated from <u>Buddleja davidii</u> Franch, by applying the dibenzoate chirality method to <u>trans</u> and <u>cis</u> dibenzoate derivatives, <u>65</u> and <u>66</u>. In the present nine-membered ring system, relative configurations were established by ^{1}H NMR coupling constants and X-ray crystallographic studies.

<u>Table 3-36.</u>

<u>67</u> $\oplus$

261 (+43.8)

242 (-7.3)

CD (48)

X ray
relative
(48)

This is a microbial product[48] produced by <u>Stachybotrys complement nov. sp.</u> K-76. Although the relative configuration was established by X-ray crystallo-

graphic studies, the absolute configuration remained unknown because of the
unavailability of suitable crystals for the X-ray Bijvoet method.   Therefore,
the absolute configuration was determined by the dibenzoate chirality method.
In this case, the observed CD spectrum of 67 was affected by the interaction
between the benzoate and the skeletal chromophore.   Here again, the
p-dimethylamionobenzoate group probably would have led to less interaction
with the natural chromophore.

Table 3-37.

370 (39,000)	370 (+4.2)	
351 (31,000)	350 (+5.6)	
	322 (−77.9)	
316 (49,000)		CD (49)
295 (45,000)	292 (+73.7)	
CHCl$_3$	CHCl$_3$	

A more involved application of the dibenzoate chirality method is exemplified by the determination[49] of the absolute configuration of trans-7,8-dihy-
droxy-7,8-dihydrobenzo[a]pyrene (71), a carcinogenic metabolite of benzo[a]py-
rene (70), and related derivatives.   The UV spectrum of bis(p-dimethylaminobenzoate) 68 shows a conspicuous intramolecular charge transfer band of the
benzoate chromophore located between the pyrenoid bands at 284–295 nm (longitudinal, see Figure 3-25) and at 334–370 nm (transverse); the p-dimethylamino
substituent was chosen so as to minimize the interaction between the dibenzoate and pyrenoid transitions.

The CD Cotton effects of glycol 71 are weak in comparison with its diben-
zoate 68, which shows strong peaks at 322 nm ($\Delta\varepsilon$ −77.9) and 292 nm ($\Delta\varepsilon$ +73.7).
Although the difference in the CD spectra above 340 nm indicates the weak
interaction of pyrenoid chromophore with the benzoate groups, it can be safely
assumed that the strong intensities and locations of the extrema at 322 and
292 nm arise mainly from the coupling between the two benzoate groups.
The ^{1}H NMR coupling constant between 7-H and 8-H is 8 Hz in CDCl$_3$, indicating that the flexible terminal ring adopts the conformation depicted in

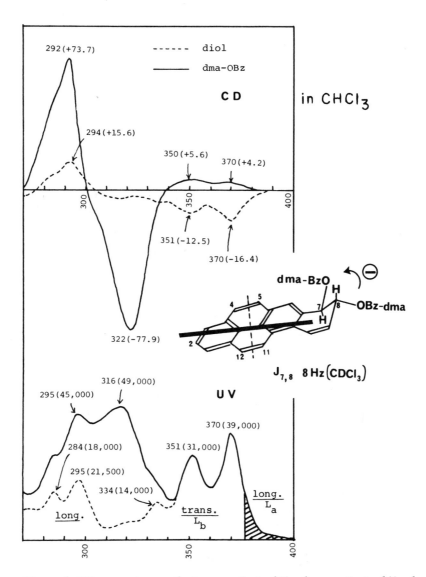

Figure 3-25. _CD and UV spectra of_ _trans-7,8-dihydroxy-7,8-dihydrobenzo[a]py-_
rene and its (p-dimethylaminobenzoate) in CHCl₃. Only regions above 270 nm
are shown due to difficulty in CD measurements. The absorptions, centered at
295 and 351 nm, are due to the longitudinal and transverse transitions of the
dihydrobenzo[a]pyrene moiety, respectively. [Adapted from reference 49].

Figure 3-26. In order to retain the conformation in the NMR solvent, the CD
spectrum of the dibenzoate was measured in a chloroform solution. Molecular
models show that in this conformation, there is little or no interaction
between the pyrenoid transitions (longitudinal and transverse) and the dibenzoate transitions because they either lie in close-to-parallel planes, i.e.,
longitudinal and 8-benzoate transitions, or because the two transitions inter-
sect (lie in the same plane), i.e., longitudinal and 7-benzoate transitions.
Thus, the CD splitting is interpreted as being caused preponderantly by the
coupling between the two benzoate chromophores, which is negative. This led
to the (7R,8R) absolute configuration of glycol 71, which had an $[\alpha]_D$
value of -460°. Glycol 71 and its enantiomer, which were resolved through
their (-)-menthoxyacetates, were separately oxidized to diol epoxide 73 (and
its enantiomer), and was incubated with poly(G). The guanosine adduct derived
from glycol 71 — i.e. adduct 74 — had the same high-performance LC reten-
tion time as the natural adduct formed by incubating benzo[a]pyrene with
bovine bronchial explants. This led to the first full structural clarifica-
tion of the adduct formed between carcinogenic hydrocarbons and mammalian RNA.

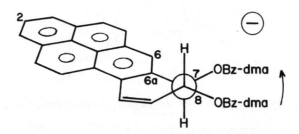

*Figure 3-26. Negative exciton chirality in the system of trans-7,8-dihydroxy-
7,8-dihydrobenzo[a]pyrene bis(p-dimethylaminobenzoate).*

Table 3-38.

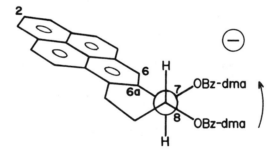	347 (45,000)	346 (+33.0)	
		333 (+20.5)	
		321 (-58.0)	
	316 (70,000)		CD (49),(50)
		295 (+19.3)	
	281 (59,000)	279 (+41.0)	
69	CHCl$_3$	CHCl$_3$	

In the case of tetrahydrodiol bis(p-dimethylaminobenzoate) 69, the situation is more complicated than that of the dihydrodiol derivative 68. The pyrenoid transverve transition is located at 316-347 nm, partially overlapping with the bis(p-dimethylaminobenzoate) transition (316 nm). Moreover, the pyrenoid longitudinal transition at 281 nm is greatly intensified in comparison with that of compound 68 (Figure 3-28). Therefore, in the case of compound 69, the two benzoates interact more extensively with the pyrenoid chro-

Figure 3-27.   Negative exciton chirality in the system of trans-7,8-*dihydroxy-7,8,9,10-tetrahydrobenzo[*a*]pyrene bis(*p*-dimethylaminobenzoate).*

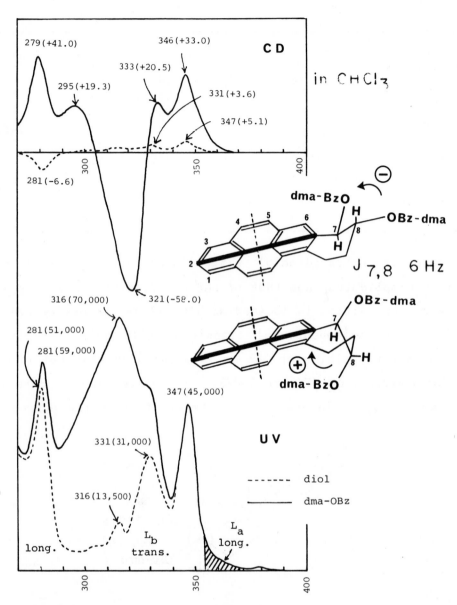

Figure 3-28. *CD and UV spectra of trans-7,8-dihydroxy-7,8,9,10-tetrahydrobenzo[a]pyrene and its (p-dimethylaminobenzoate) in CHCl₃. [Adapted from reference 49].*

mophore as compared to 68. The observed CD spectrum is complex; both of the pyrenoid transitions exhibit strong Cotton effects (λ_{ext} 346 nm, $\Delta\varepsilon$ +33.0; 279 nm, $\Delta\varepsilon$ +41.0). It is therefore difficult to rigorously assign the CD Cotton effects. However, comparison of the spectrum with that of compound 68 leads to the conclusion that the extrema at 321 and 295 nm are due to exciton split Cotton effects arising from the dibenzoate moiety. Since the observed Cotton effects are of negative chirality, the absolute configuration is deduced to be (7R,8R); the conclusion is the same as that derived from Figure 3-25.

Table 3-39.

326 (+52.0)	
316 (0.0)	CD (50)
305 (-70.8)	
	MeOH

75

Koreeda, et al.[50] independently established the absolute stereochemistry of benzo[a]pyrene metabolites by applying the benzoate chirality method. In their studies, trans-7,8-dihydroxy-4,5,7,8,9,10,11,12-octahydrobenzo[a] pyrene bis(p-dimethylaminobenzoate) (75) was chosen in order to diminish the contribution of the skeletal chromophore which disturbs the observation of the split CD curve due to the vicinal dibenzoate. Namely, absence of the pyrenoid transition around 340 nm in the CD spectrum of the bis(p-dimethylaminobenzoate) <u>75</u> allows one to arrive at an unequivocal assignment of the positive exciton chirality between the two benzoate groups, and this establishes the (7<u>S</u>,8<u>S</u>) absolute configuration of compound <u>75</u>.

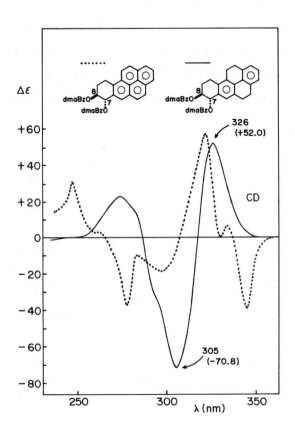

Figure 3-29. *CD spectra of bis(p-dimethylaminobenzoates) of* <u>trans-7,8-dihydroxy-7,8,9,10-tetrahydrobenzo[a]pyrene</u> *and* <u>trans-7,8-dihydroxy-4,5,7,8,9,10, 11,12-octahydrobenzo[a]pyrene</u> *in MeOH. The CD curve of the tetrahydrodiol bis(p-dimethylaminobenzoate)* <u>69</u> *in this figure is slightly different from that in Figure 3-28; this is presumably due to a solvent effect. [Adapted from reference 50.]*

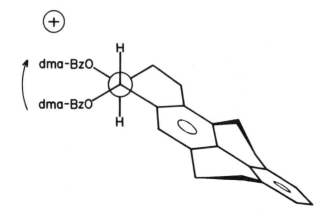

Figure 3-30. Positive exciton chirality in the system of trans-7,8-dihydroxy-
4,5,7,8,9,10,11,12-octahydrobenzo[a]pyrene bis(p-dimethylaminobenzoate), where
the tetrahydropyrene moiety is tentatively depicted in one distorted conformation.   It is possible that the system is distorted in the opposite direction
but this does not affect the chirality of the vicinal dibenzoate group.

A striking example proving the nonempirical nature and utility of the CD
exciton chirality method is the reversal[54] of the absolute configuration of
clerodin (78),[51] a basic compound of clerodane diterpenes.   The absolute
stereochemistries of clerodin and related insect antifeedants, caryoptin (79)
and 3-epicaryoptin (80) had been previously determined by the X-ray Bijvoet
method[52] and by the chiroptical method,[53] and had been believed to be the
mirror image of formulas 78, 79, and 80, respectively, for a long time.
However, the recent application[54] of the CD exciton chirality method led
to the conclusion that the X-ray results were wrong and that the absolute
stereochemistries should be revised and represented by formulas 78, 79, and
80, respectively.

clerodin	caryoptin	3-epicaryoptin
78	79	80

As described in the previous sections, it is well established that the absolute stereochemistries as determined by the X-ray Bijvoet and CD exciton methods are consistent with each other. In the case of bis(p-chlorobenzoates) of 3-epicaryoptin and caryoptin derivatives, (76) and (77), however, Munakata and coworkers[53] reported that the absolute configuration determined by the CD exciton chirality method disagreed with that derived from the X-ray results of clerodin. In order to account for this discrepancy, they assumed that the conformation of the benzoate group was twisted by a seven-membered intramolecular hydrogen bonding. Thus, they believed the X-ray results and abandoned

Table 3-40.

		CD (53)
$\quad\oplus\quad$	*240 (37,200)*	
		247.5 (+17.8) CD exciton (54)
		230 (-9.2) $\Delta[\phi]_D$ (54)
	EtOH	20% dioxane/ X ray (58) EtOH

76

		CD (53)
	247.5 (-28.8)	CD exciton
240 (20,900)		(54)
	230 (+7.5)	
		$\Delta[\phi]_D$ (54)
EtOH	20% dioxane/	
	EtOH	Correl. (53)

_77_

the exciton CD data, and concluded that these two compounds were exceptions of the CD exciton chirality method.

On the other hand, they also determined the absolute configuration of clerodendrin A (81) with the same clerodane skeleton by X-ray[55,57] and by chemical correlation.[55,56] If all of these results are true, it is noteworthy that clerodin, caryoptin, and 3-epicaryoptin are antipodal to clerodendrin A in all corresponding chiral centers, in spite of isolation from the plants of same genus, Clerodendron. Therefore, it was of interest to check whether the benzoate conformation was indeed twisted by intramolecular hydrogen bonding, and to determine the absolute configuration of clerodin, caryoptin, and 3-epi-caryoptin for biosynthetic correlations.

clerodendrin A
81

(R)-(−)-2-hydroxy-
2-methylbutyric acid

Table 3-41.

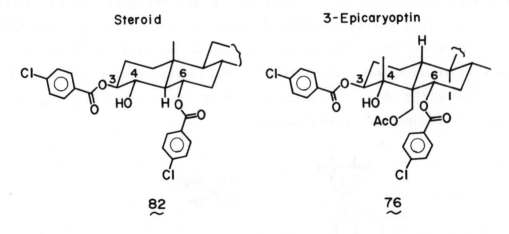

	239.3 (33,500)	246.2 (+27.0) 237.0 (0.0) 231.0 (-13.8)	CD (54)
			X ray (3)
	EtOH	EtOH	

As depicted in Figure 3-31, 5α-cholestane-3β,4α,6α-triol 3,6-bis(p-chlorobenzoate) (82) is a suitable model compound for examining whether the benzoate group in 3-epicaryoptin derivative 76 really has a twisted conformation caused by intramolecular hydrogen bonding. The CD spectrum of the steroidal dibenzoate 82[54] exhibits typical exciton split Cotton effects

Steroid   **3-Epicaryoptin**

82   76

Figure 3-31. 5α-Cholestane-3β,4α,6α-triol 3,6-bis(p-chlorobenzoate) and 3-epicaryoptin derivative 3,6-bis(p-chlorobenzoate) have identical chirality in principal chiral centers, i.e., 3, 4, and 6 positions.

of positive first and negative second signs, λ_{ext} 246.2 nm, $\Delta\varepsilon$ +27.0 and 231.0 nm, $\Delta\varepsilon$ -13.8 (EtOH), corresponding to the normal positive exciton chirality between the two benzoate groups (Figure 3-32). It is therefore clear that the exciton chirality between the two benzoate groups in question is not reversed by the adjacent hydroxyl group.

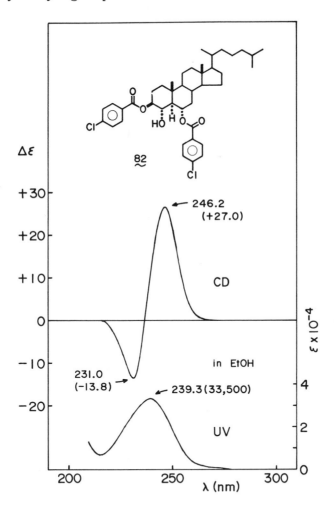

Figure 3-32. CD and UV spectra of 5α-cholestane-3β,4α,6α-triol 3,6-bis(p-chlorobenzoate).

Table 3-42.

239.3 (38,800)	247.5 (+28.8) 237.6 (0.0) 230.0 (-13.9)	CD (54) X ray (3)
2% dioxane/ EtOH	5% dioxane/ EtOH	

Table 3-43.

240.5 (35,500)	295.0 (-1.4) 263.0 (-1.0) 247.0 (+17.4) 237.0 (0.0) 230.0 (-7.6)	CD (54) X ray (3)
EtOH	3% dioxane/ EtOH	

240.2 (39,200)	248.5 (+26.2) 238.8 (0.0) 231.0 (-15.5)	CD (54) X ray (3)
1% dioxane/ EtOH	2% dioxane/ EtOH	

The above fact is also supported by the following CD data.   Dibenzoate 83, having no hydroxyl group, shows CD Cotton effects of the same sign and similar amplitude, λ_{ext} 247.5 nm, $\Delta\epsilon$ +28.8 and λ_{ext} 230.0 nm, $\Delta\epsilon$ -13.9 (5% dioxane in EtOH)(Figure 3-33).   Moreover, compounds 84 and 85 also exhibit positive first and negative second Cotton effects, as shown in Table 3-43.

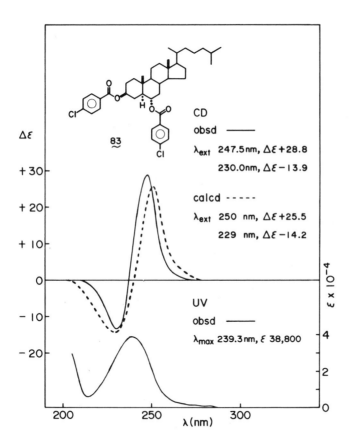

Figure 3-33.   CD and UV spectra of 5α-cholestane-3β,6α-diol bis(p-chlorobenzoate).   Dotted line is the theoretically calculated CD curve based on the chiral exciton coupling mechanism.

Since the reported CD Cotton effects of 3-epicaryoptin derivative dibenzoate 76 are of positive exciton chirality, λ_{ext} 247.5 nm, $\Delta\varepsilon$ +17.8 and 230 nm, $\Delta\varepsilon$ −9.2 (20% dioxane in EtOH), comparison of the data of 82 and 76 leads to the conclusion that the absolute configuration of 3-epicaryoptin should be revised and expressed by formula 80. The same is true for caryoptin; since the reported CD Cotton effects of 77 are λ_{ext} 247.5 nm, $\Delta\varepsilon$ −28.8 and 230 nm, $\Delta\varepsilon$ +7.5 (20% dioxane in EtOH), the absolute configuration of caryoptin should be represented by formula 79.

The above conclusion was confirmed by the following molecular rotation data.[54] Catalytic hydrogenation of the heterocyclic ring of clerodin, caryoptin, and 3-epicaryoptin derivatives yielded the dihydro derivatives. The molecular rotation difference between dihydro and unsaturated compounds, $\Delta[\phi]_D = [\phi]_D$(dihydro) − $[\phi]_D$(unsaturated), is always positive and almost constant (about +130°) throughout the series of clerodin, caryoptin, and 3-epi-caryoptin, as exemplified in Figure 3-34. Thus, the molecular rotation difference $\Delta[\phi]_D$ is mainly due to the chiroptical change of the furo-furan ring, the contribution of the trans-decalin moiety being negligible. In the case of clerodendrin A derivatives, $\Delta[\phi]_D$ is also always positive and almost constant (about +150°). Similar values of $\Delta[\phi]_D$ are observed in the case of lactone formation. Since the relative stereochemistries of the furo-furan ring and other principal chiral centers of clerodin, caryoptin, 3-epicaryoptin and clerodendrin A are identical with each other, the rotation data strongly indicate that all of these compounds have the same absolute configuration. Moreover, since the absolute configuration of clerodendrin A had been doubly established by the X-ray Bijvoet and chemical correlation methods, the absolute stereochemistry of clerodin, caryoptin, and 3-epicaryoptin should be represented by formulas 78, 79, and 80, respectively.[54]

clerodin 78

$[\alpha]_D = -47° (c\,1.66, \text{CHCl}_3)$

$[\phi]_D = -204.2°$

dihydroclerodin 86

$[\alpha]_D = -20° (c\,1.44, \text{CHCl}_3)$

$[\phi]_D = -87.3°$

$\Delta[\phi]_D = -87.3° - (-204.2°) = +116.9°$

clerodendrin A 81

$[\alpha]_D = +7.4° (c\,3.25, \text{CHCl}_3)$

$[\phi]_D = +44.9°$

87

$[\alpha]_D = +30.4° (c\,1.30, \text{CHCl}_3)$

$[\phi]_D = +185.0°$

$\Delta[\phi]_D = +185.0 - 44.9 = +140.1°$

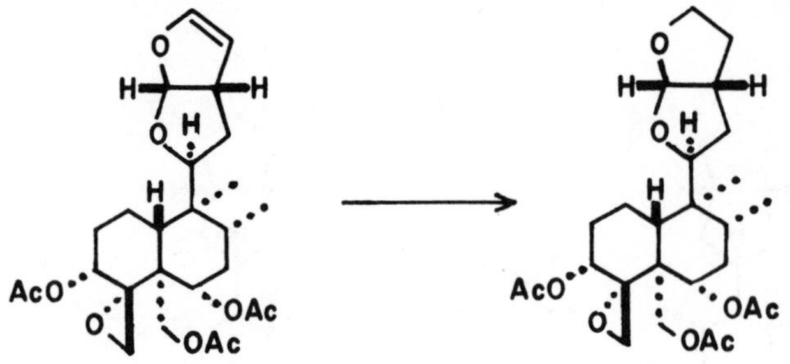

caryoptin 79

$[\alpha]_D = -91°(c\,1.10, CHCl_3)$

$[\phi]_D = -448.2°$

88

$[\alpha]_D = -63°(c\,1.13, CHCl_3)$

$[\phi]_D = -311.6°$

$\Delta[\phi]_D = +136.6°$

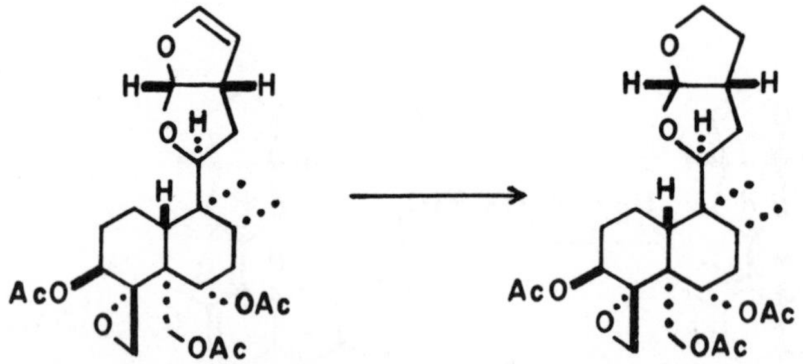

3-epicaryoptin 80

$[\alpha]_D = -70°(c\,1.01, CHCl_3)$

$[\phi]_D = -344.8°$

89

$[\alpha]_D = -42°(c\,1.04, CHCl_3)$

$[\phi]_D = -207.7°$

$\Delta[\phi]_D = -207.7° - (-344.8°) = +137.1°$

Figure 3-34. *Clerodin, clerodendrin A, caryoptin, and 3-epicaryoptin all exhibit positive shifts in D-line molecular rotations ($[\phi]_D$) when catalytically hydrogenated.*

At the same time, Rogers, et al.[58] independently arrived at the same conclusion by X-ray studies of 3-epicaryoptin; Sim,[58] one of the authors of the original paper on the X-ray studies of clerodin, revised the previous absolute configurational assignment of clerodin.

Table 3-44.

90	(+)	243 (35,000) MeOH	252 (+24.8) 235 (-11.6)	CD (59) X ray (58)
91	(+)	245 (20,000) MeOH	252 (+18.1) 236 (-7.2)	CD (59) X ray (58)

It was also shown that blocking of the hydroxyl group by urethane derivatization did not change the sign of the exciton split Cotton effects,[59] as indicated in Table 3-44. Therefore, it is evident that an intramolecular hydrogen bond, if any, was not involved in the factors governing the exciton chirality between the two benzoate moieties.

The following examples show that the dibenzoate chirality method is applicable to flexible compounds such as macrolides or acyclic compounds, provided the relative conformation of glycol moiety is assignable from other physical or chemical data; e.g., NMR coupling constants, etc.

Table 3-45.

	274 (2,000)	*237 (-21.6)*	
	229 (36,300)		CD (60)
		213 (-11.3)	
92	MeOH	MeOH	

For example, MacMillan, et al.[60] determined the absolute stereochemistry of colletodiol, a metabolite of <u>Colletotrichum capsici</u>, by application of the present method in conjunction with NMR and chemical correlation data. The dibenzoate <u>92</u> of colletodiol exhibited a negative first Cotton effect and a positive second Cotton effect overlapping with the negative Cotton effect of two conjugated lactone moieties. Subtraction of the CD curve of the diacetate from that of the dibenzoate clearly gave a split CD Cotton effect with negative chirality which established the absolute stereochemistry of the glycol moiety (Figure 3-35). A <u>para</u>-substituted benzoate chromophore would have been more suitable to avoid interference from the two unsaturated lactone chromophores.

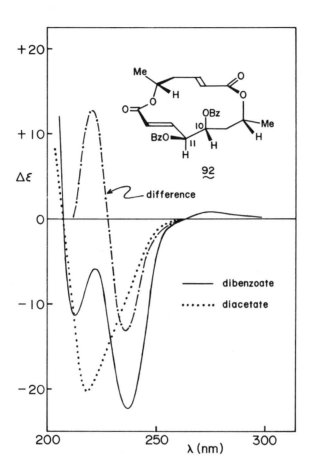

Figure 3-35. CD spectra of colletodiol dibenzoate and diacetate.   The difference curve exhibits negative first and positive second Cotton effects. [Adapted from reference 60.]

McGahren and coworkers[61] established the absolute configuration of the acyclic glycol moiety of a fungal metabolite, 2'-hydroxypestalotin (94), isolated from Penicillium sp., by application of this method. Namely, a threo configuration of the glycol part was deduced on the basis that the 1,2-thiocarbonate 95 afforded a trans olefin 96 by cis elimination; in addition, the 6.0 Hz ^{1}H NMR coupling constant of the glycol part in the dianisoate 93 led to a trans relation for the vicinal hydrogens. This led to two possible isomers having the conformations depicted in Figure 3-36. On the other hand, the CD spectrum of the dianisoate 93 showed two strong split Cotton effects, λ_{ext} 267.5 nm, $\Delta\varepsilon$ −19.0 and 250 nm, $\Delta\varepsilon$ +6.7, indicating a negative chirality for the glycol moiety. The absolute configuration of compound 93 was thus

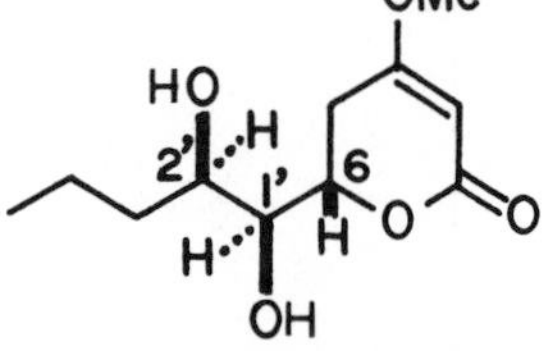

(6S,1'S,2'R)

94

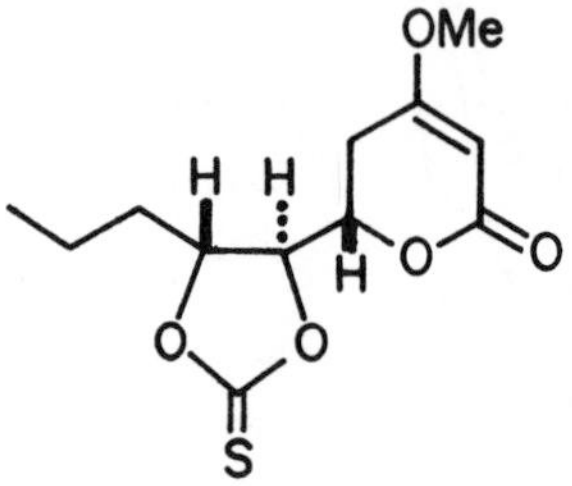

95

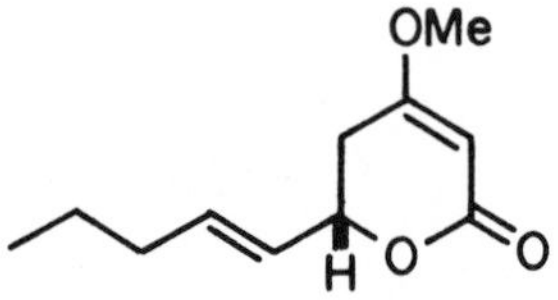

96

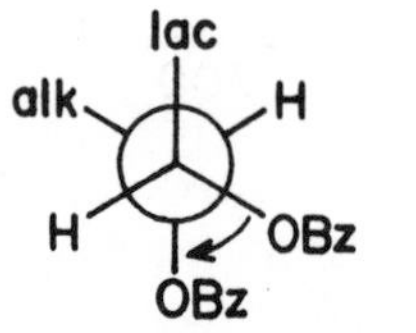

(6S,1'R,2'S) positive

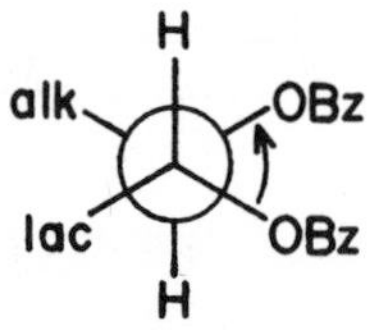

(6S,1'S,2'R) negative

Figure 3-36. Stereochemistry and conformation of 2'-hydroxypestalotin.

Table 3-46.

93	

267.5 (-19.0)
                    CD (61)
250   (+6.7)
219   (-4.9)   Correl.
211   (+9.8)   (62),(63)

MeOH

elucidated as depicted.   The assignment was later confirmed by syntheses of the optically active pestalotin.[62,63]

In the application of the dibenzoate chirality method to determine the absolute configurations of acyclic or flexible compounds, it is important that the conformation be established without ambiguity.   Namely, success in the assignment of absolute stereochemistry by the present method depends on the reliability of the conformation deduced by other physical or chemical methods.

Table 3-47.

R = [benzoyl-Cl structure]

97

246 (+17.2)
                    CD (64)
229 (-7.2)

MeOH

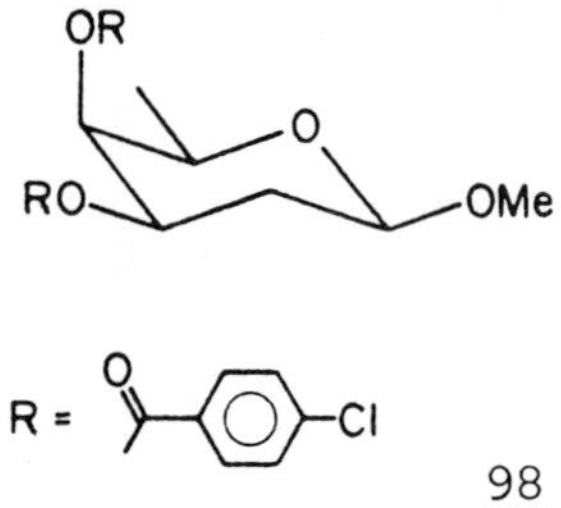

246 (+28.9)

229 (-7.5)

CD (64)

MeOH

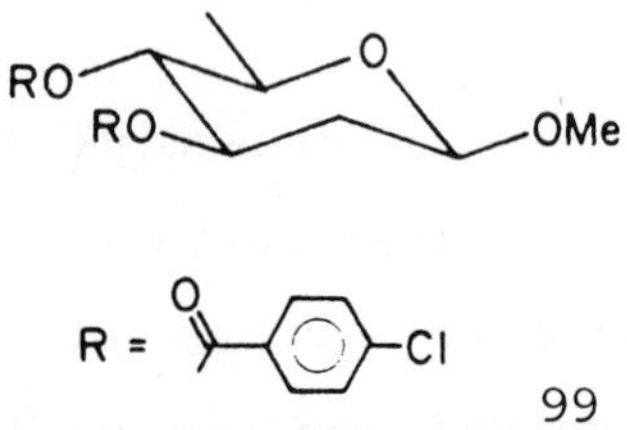

245 (−29.0)

230 (+14.5)

CD (64)

MeOH

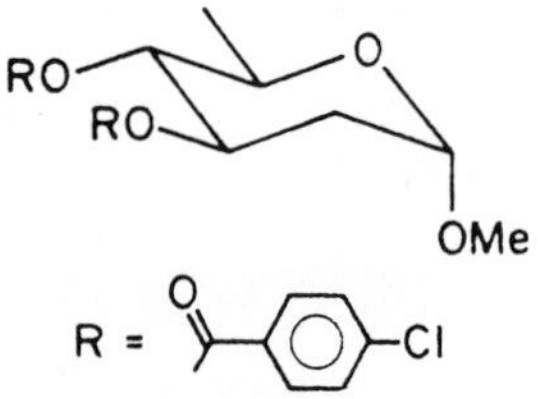

246 (−37.1)

229 (+17.1)

CD (64)

MeOH

247 (-10.5)

CD (64)

MeOH

Applications of the dibenzoate chirality method to pyranose and furanose diols permits assignment of absolute configuration or determinations of preferred conformations;[64] it also reflects subtle conformational distortions caused by steric interactions. For example, pyranose bis(p-chlorobenzoates) 97, 98, 99, and 100 show typical split Cotton effects, the signs of which are in agreement with predictions without exceptions. As discussed in section 3-1-C, when the dihedral angle between the benzoate groups is 180°, theoretically no interaction should occur. The CD spectrum of diaxial dibenzoate 101 exhibits only a single Cotton effect at the perturbed wavelength of 247 nm ($\Delta\varepsilon$ -10.5). The wavelength and negative sign of the relatively weak Cotton effect show that the diaxial benzoate groups are distorted towards the outside of the ring to alleviate 1,3-diaxial interaction.

Table 3-48.

102

229 (25,600)   *235 (+9.3)*   CD (4)

EtOH   EtOH

103

234 (-10.6)   CD (4)
224 (0.0)
217 (+3.5)

EtOH

In the case of furanose systems,[4] the anomeric riboside dibenzoates (102) and (103) show coupled Cotton effects of opposite chiralities indicating inverse conformations of the five-membered riboside ring. Namely, compound 102 adopts a conformation with positive chirality, while compound 103 adopts a conformation with negative chirality (Figure 3-37). As exemplified above, five-membered rings are flexible; the CD data of cis-glycol dibenzoates of isolated five-membered rings thus indicate the absolute conformation rather than configuration.

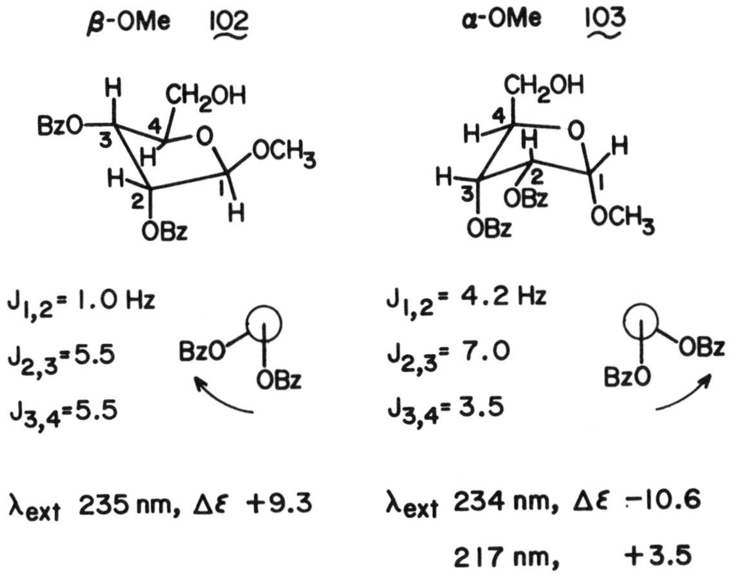

β-OMe 102   α-OMe 103

$J_{1,2}= 1.0$ Hz $J_{1,2}= 4.2$ Hz

$J_{2,3}=5.5$ $J_{2,3}= 7.0$

$J_{3,4}=5.5$ $J_{3,4}= 3.5$

λ_{ext} 235 nm, $\Delta\varepsilon$ +9.3 λ_{ext} 234 nm, $\Delta\varepsilon$ −10.6

217 nm,   +3.5

Figure 3-37. Conformation and exicton chirality of anomeric riboside dibenzoates.

Table 3-49.

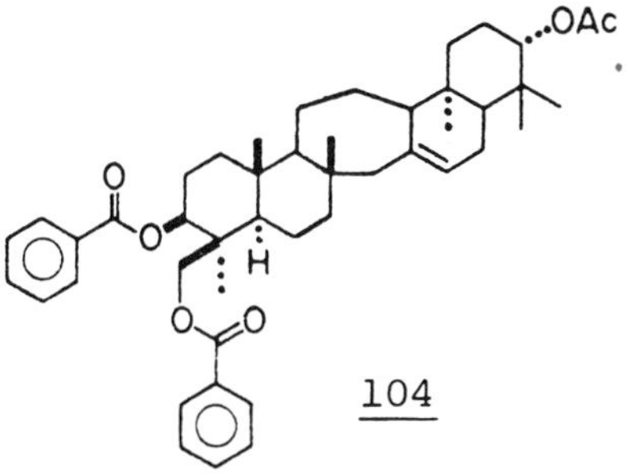

229 (25,300)	236 (+14.9)	CD (4)
	226 (0.0)	
	212 (−10.8)	
10% dioxane/ EtOH	10% dioxane/ EtOH	

$$233 \; (+14.0)$$
$$229.8 \; (24,700) \qquad 222 \; (\;\; 0.0) \qquad \text{CD (4)}$$
$$219 \; (\; -1.9)$$

EtOH   EtOH

The situation becomes more complicated in systems containing primary alcohol groups such as compounds 104 and 105.[4]   Both triterpenoid dibenzoates exhibit Cotton effects of positive chirality, irrespective of the difference in the absolute configurations of the primary benzoate groups.   Therefore, although it is difficult to determine the absolute configuration from the present data, the CD data inversely permits deduction of the absolute conformation of the dibenzoate moieties; in the present cases, the two dibenzoates adopt conformations with positive chirality.

Table 3-50.

$$253.0 \; (+35.4)$$
$$245 \; (38,000) \qquad 242.5 \; (\;\; 0.0) \qquad \text{CD (65)}$$
$$237 \quad (-11.7)$$

dioxane   dioxane

			252.5 (-28.4)	
	245 (39,000)	$\ominus$	242.2 (0.0)	CD (65)
			234.5 (+7.9)	
	dioxane		dioxane	

107

			253.5 (+27.7)	
	245 (38,000)	$\oplus$	242.7 (0.0)	CD (65)
			237 (-8.7)	
	dioxane		dioxane	

108

Kato and coworker[65] reported the conformational analyses of furo-furan ring compounds, e.g., 1,4:3,6-dianhydro-D-mannitol (106), -D-sorbitol
(107), and -L-iditol (108), by application of the exciton chirality method in
conjunction with NMR studies.   The resemblance in the ^{1}H NMR coupling
constants of the free glycol and its bis(p-bromobenzoate) indicate that the
conformations are similar.   Although these compounds are 1,4-dibenzoate,
coupled Cotton effects are clearly observable.   In compound 106, the exciton
chirality between the two benzoate groups is always positive irrespective of
conformational mobility of the furo-furan ring, as depicted in Figure 3-38.
The prediction is satisfied by the observed CD data as indicated in Table 3-50.
Similarly, the chirality of compound 107 is always negative.   On the other

hand, in the case of compound <u>108</u>, the chirality between the two benzoate groups could not be definitely predicted when the dibenzoate adopts a conformation with undistorted five-membered rings. However, the CD data indicate that the furo-furan ring is distorted in such a way that the two benzoate groups are positively twisted. The results are also reasonable from examination of molecular models.

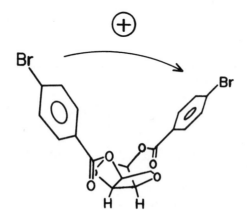

Figure 3-38. *Positive exciton chirality in 1,4;3,6-dianhydro-D-mannitol bis(p-bromobenzoate).*

3-3-B.   Synthetic Organic Chiral Compounds

The dibenzoate chirality method is also useful for determining the absolute configuration of synthetic organic compounds, as exemplified by the following compounds.

Table 3-51.

		321	(-36.6)	
		307	(0.0)	CD (67)
	310.5 (49,000)	295	(+15.7)	
		247	(-0.7)	
	229.0 (12,900)	227.5	(+3.4)	

(-)   **109**

EtOH   EtOH

252.0	(-19.9)	
241.0	(0.0)	CD (5)
234.5	(+6.7)	
211.5	(+1.6)	

(-)   **110**

EtOH

Spiro[4.4]nonane-1,6-dione (**111**) and -1,6-diene are typical examples of chiral spirans which behave theoretically interesting nonbonding homoconjugation. Their absolute configurations were derived[66] on the basis of the empirical Horeau's method which, in spite of its empirical nature, leads to correct absolute configurations in the majority of cases. The dibenzoate chirality method was applied[67] to this spiro-system in order to check the absolute configuration by the nonempirical method. Reduction of (-)-diketone **111** afforded (-)-cis,trans-spiro[4.4]nonane-1,6-diol (**112**), the relative configuration of which was readily assignable from the lack of C_2-symmetry in the NMR spectrum (Figure 3-40). Bis(p̲-dimethylaminobenzoate) <u>109</u> showed typical split Cotton effects of negative chirality establishing the (1S̲,5S̲,6R̲) absolute configuration which was in agreement with that obtained by the Horeau's method (Figure 3-39). The absolute stereochemistry of spirans is thus easily and unambiguously determinable by introducing p̲-dimethylaminobenzoate group as reporting chromophores.

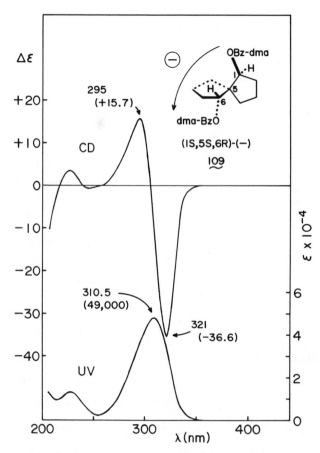

Figure 3-39. CD and UV spectra of (1S̲,5S̲,6R̲)-(-)-spiro[4.4]nonane-1, 6-diol bis(p̲-dimethylaminobenzoate). [Adapted from reference 67.]

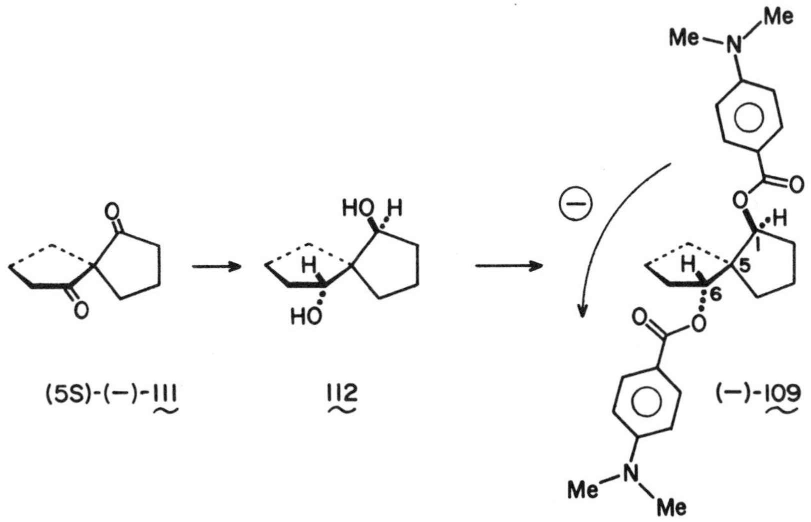

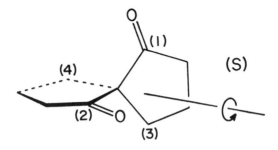

(5S)-(−)-**III**   **II2**   (−)-**I09**

Figure 3-40.   Chemical correlation of (5S)-(−)-spiro[4.4]nonane-1,6-dione with (−)-cis,trans-spiro[4.4]nonane-1,6-diol bis(p-dimethylaminobenzoate).

The (R) and (S) nomenclature of spiro compounds with central chirality is defined by applying the sequence rule as follows:[68] in the case of spiro [4.4]nonane-1,6-dione, shown in Figure 3-41, either of the carbonyl carbon atoms is given the highest priority (1). The other carbonyl carbon atom is given the second priority (2). Of the

Figure 3-41.   (S) configuration of spiro[4.4]nonane-1,6-dione.

methylene carbon atoms attached to the spiro center, that which occupies the same ring as the most preferred carbonyl carbon atom is assigned priority (3).   The remaining one then has priority (4). The sequence rule path is counterclockwise and hence (S).

Table 3-52.

	323.5 (-45.4)	
310.9 (55,100)	310.2 (0.0)	CD (81a)
	298.0 (+22.3)	
	250.0 (-0.8)	
	222.5 (+7.5)	

113   EtOH   EtOH

The absolute configuration of 2,2'-spirobiindanes[81a] was determined in a similar manner.   As illustrated in Figure 3-42, the cis,trans-2,2'-spirobiindan-1,1'-diol bis(p-dimethylaminobenzoate) (113) exhibits strong split Cotton effects of negative chirality; this leads to the (1R,1'S,2S) absolute configuration. Bis(p-chlorobenzoate) also shows split CD but the amplitude is smaller than that of bis(p-dimethylaminobenzoate) 113 (dotted line in Figure 3-42).   The p-dimethylaminobenzoate chromophore is thus more favorable.   The absolute configuration determined is in agreement with that obtained by chemical correlation and NMR studies of a metal complex.[81b]

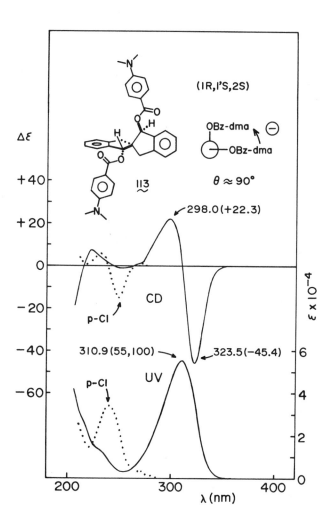

Figure 3-42. CD and UV spectra of *(1S,1'R,2S)-(-)-2,2'-spirobiindan-1,1'-diol bis(p-dimethylaminobenzoate) in ethanol. The dotted lines are those of bis(p-chlorobenzoate).*

On the other hand, <u>trans,trans</u>-2,2'-spirobiindan-1,1'-diol bis(<u>p</u>-dimethylaminobenzoate) (<u>114</u>) merely shows a single negative Cotton effect at 304 nm (Figure 3-43). This is due to the fact that the dihedral angle between the two benzoate chromophores is approximately 180°; the exciton chirality between them is hence almost nil.

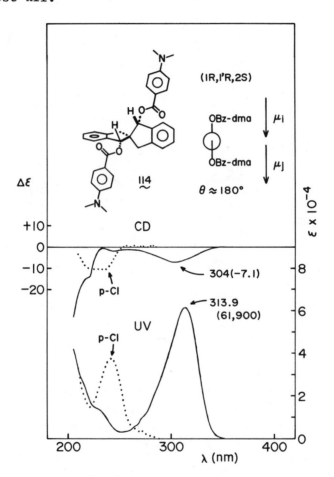

Figure 3-43. CD *and* UV *spectra of* (1<u>R</u>,1'<u>R</u>,2<u>S</u>)-2,2'-*spirobiindan-1,1'-diol* bis(<u>p</u>-*dimethylaminobenzoate) in ethanol. The dotted curves are those of bis* (<u>p</u>-*chlorobenzoate).*

Table 3-53.

	236 (+10.0)	CD (69)
	221 (-3.2)	Correl. (69)
115	EtOH	

Tichy[69] revised the absolute configuration of twistane on the basis of chemical correlation and the CD Cotton effects of the positive chirality observed for the twistane-diol dibenzoate (115). The two methods led to the same conclusion and hence to the revised absolute configuration.

Table 3-54.

	CD (70)	
	Correl. (70)	
116		

Nakazaki et al.[70] determined the absolute configuration of (+)-trans-1,2-dihydroxyacenaphthene by means of chemical degradation and application of the dibenzoate chirality method to dibenzoate 116 with a dihedral angle of 120°. A <u>para</u>-substituted benzoate chromophore would have been preferable to avoid interference from the naphthalene chromophore.

Table 3-55.

_234 (14,400)_   _237 (-17.5)_   CD (71)

_221 (+7.6)_

MeOH

<u>117</u>

The absolute configuration of (−)-adamantane-diol was also determined by application of the exciton chirality method.[71]   Namely, the 1,5-dibenzoate system <u>117</u> still exhibits a relatively strong split Cotton effect, λ_{ext} 237 nm, $\Delta\varepsilon$ −17.5 and 221 nm, $\Delta\varepsilon$ +7.6, and this led to the absolute configuration of negative chirality as depicted.

3-4.   Dibenzoate Chirality Method for Amino Alcohol and Diamine Systems

3-4-A.   Amino Alcohols

As described in section 2-3, the intramolecular charge transfer transition or ^1L_a transition of the benzamide chromophore is also suitable for chiral exciton coupling.

Table 3-56.

270sh(2,000)

237 (-6.8)

227 (26,300)

222 (+3.2)   CD (72)

MeOH   MeOH

118

270sh   (900)

237 (-4.0)   CD (72)

228 (15,500)

222 (+2.8)

MeOH   MeOH

119

no strong
Cotton effect   CD (72)
above 220 nm.

MeOH

120

Johnson et al.[72] determined the absolute stereochemistry of vancosa-
mine, an amino sugar derived from the antibiotic vancomycin produced by Streptomyces orientalis, by applying the dibenzoate chirality method to the amino-
alcohol moiety of β- and α-anomers 118 and 119.   The relative configuration
and conformation of the benzamide-benzoate systems were established by NMR
spectroscopy, i.e., coupling constants and NOE (nuclear Overhauser effect)
enhancement.   From the negative and positive signs of the first and second
Cotton effects, respectively, the absolute configuration of the amino sugar
was established as depicted in Table 3-56; these results were in agreement
with the results derived from the Cupra A method.   On the other hand, mono-
benzamide 120 showed no strong Cotton effect above 220 nm.   It should be noted
that, in contrast to the difficulty in the preparation of benzoates of tertiary alcohols, the benzamide of the amino group attached to a tertiary carbon
was readily prepared and employed in the present exciton method.

Table 3-57.

121	⊖	228 (21,400)	238 (−16.9) 228 (0.0) 222 (+9.6)	CD (73)
		MeOH	MeOH	

Nagai and coworkers[73] reported the CD data of benzoate-benzamides of
cyclic and acyclic amino alcohols listed in Tables 3-57 and 3-58. The benzoate-

benzamide 121 of 2-aminocyclohexanol exhibits split Cotton effects of negative chirality, as expected from the known absolute configuration.

Table 3-58.

122	$\oplus$	227.5 (22,300) MeOH	236 (+5.5) 227 (0.0) 221 (-3.0) MeOH	CD (73)

123	$\ominus$	228.5 (23,500) MeOH	238 (-6.3) 227 (0.0) 221 (+4.2) MeOH	CD (74)

The acyclic benzoate-benzamides 122 and 123 show Cotton effects of opposite signs to each other, reflecting the difference in absolute configurations.[73]   The signs of the observed Cotton effects are explicable on the basis of the following conformational analysis.   For example, three staggered conformers are possible for 122 as depicted in Figure 3-44, where conformer (c) is unfavorable because of steric hindrance.   The remaining conformer (a) has a positive chirality, while conformer (b) has no exciton chirality because of the <u>trans</u>-conformation of the benzoate and benzamide groups.   Thus an (<u>S</u>) configuration can be assigned to the benzoate–benzamide exhibiting split Cotton effects of positive chirality.

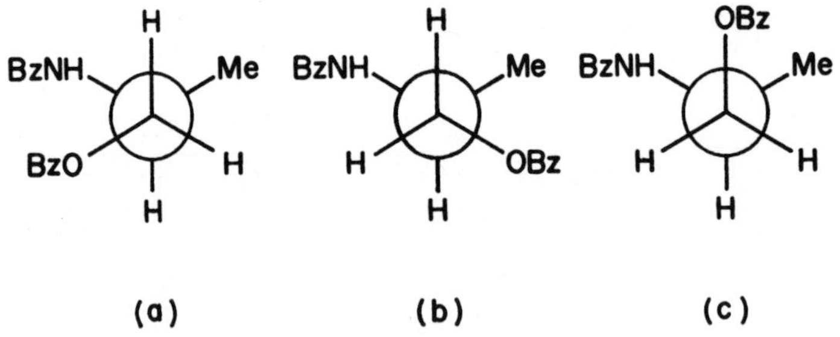

Figure 3-44.   *Three staggered conformers of benzoate–benzamide of 2-amino-1-propanol.*

3-4-B. Diamines

The exciton chirality method is also applicable to diamine systems.

Table 3-59.

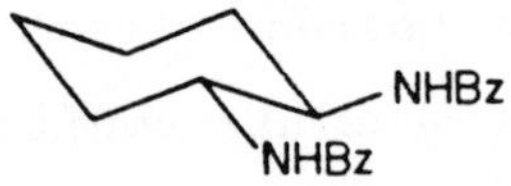	(−)	225.5 (20,000)	239 (−13.0) 227 (0.0) 220 (+9.4)	CD (73)
<u>124</u>		MeOH	MeOH	

As in the case of benzamide-benzoate of 2-aminocyclohexanol, cyclohexane-1,2-diamine dibenzamide (124) exhibits typical split Cotton effects of negative exciton chirality.[73]

Table 3-60.

		239 (-5.9)	
125	226.5 (20,900)	229 (0.0)	CD (73)
		219 (+5.3)	X ray (75)
	MeOH	MeOH	

		240 (-1.8)	
126	226.5 (21,600)	232 (0.0)	CD (73)
		222 (+2.3)	
	MeOH	MeOH	

Acyclic dibenzamides 125 and 126 show split Cotton effects, which can be accounted for in a manner similar to the case of acyclic amino alcohol benzoate-benzamides 122 and 123.

The interpretation of the CD data of acyclic benzoate-benzamides 122 and 123, and dibenzamides 125 and 126 leads to the following experimental generalization: dibenzoates of glycols, benzoate-benzamides of amino alcohols, and dibenzamides of diamines with the absolute configuration (a) depicted in Figure 3-45 exhibit exciton split Cotton effects of positive chirality, while compounds with the configuration (b) show Cotton effects of negative chirality.   This generalization, which is supported by conformational analysis, should be applicable to other natural acyclic compounds.

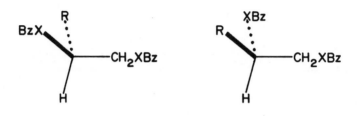

X; O or NH

Cotton effects of   Cotton effects of
positive chirality   negative chirality

(a)   (b)

Figure 3-45. *Relation between absolute configuration and sign of exciton Cotton effects in the systems of acyclic vicinal dibenzoate, benzoate-benzamide, and dibenzamide, where R is alkyl or carboxylic acid group.*

Table 3-61.

HOOC NHBz H NHBz 127	227.5 (21,900) MeOH	240 (-2.2) 231 (0.0) 223 (+2.2) MeOH	CD (73)
HOOC NHBz H NHBz 128	226.5 (22,000) MeOH	241 (-1.0) 233 (0.0) 224 (+1.3) MeOH	CD (73)
HOOC NHBz H NHBz 129	226.5 (21,200) MeOH	($\Delta\varepsilon < -0.2$) ($\Delta\varepsilon < +0.2$) MeOH	CD (73)

In the case of remote diamine dibenzamides <u>127</u>, <u>128</u>, and <u>129</u>, the sign of split Cotton effects still reflects the absolute configuration;[73] the $\Delta\varepsilon$ value decreases with increment of the interchromophoric distance between two benzamide chromophores.

3-5.   The Exciton Chirality Method for Tribenzoate Systems

<u>3-5-A.   Sugar Tribenzoates</u>[64,78]

The exciton chirality method that has been successfully applied to dibenzoate systems is also applicable to tribenzoate systems.   In these cases, the following aspects are significant.   Namely, if the 1,2,3-triol moiety is such that the chiralities between the 1,2- 2,3-, and 1,3-diol groups are all negative or positive, the first Cotton effect amplitude is greatly augmented and has large $\Delta\varepsilon$ values of minus or plus ca. 80, respectively.

Table 3-62.

(+) (2-3)	
(+) (3-4)	246 (+81.3)
(+) (2-4)	235 (0.0) CD (64)
	230 (-25.6)
	hexane

<u>130</u>

$R = $ p-chlorobenzoyl

For example, a sugar tribenzoate, α-methyl-L-arabinopyranoside 2,3,4-tris (p-chlorobenzoate) (<u>130</u>), has a 1,2,3-tribenzoate moiety, in which the two 1,2- and one 1,3-dibenzoate moieties all are of positive chirality (Figure 3-46). The CD spectrum shows a positive first Cotton effect at 246 nm, with the large value of Δε +81.3, and a negative Cotton effect at 230 nm, Δε −25.6 (Figure 3-47). The same is true for other sugar tribenzoates <u>131</u>, <u>132</u>, and <u>133</u> (Figure 3-47).

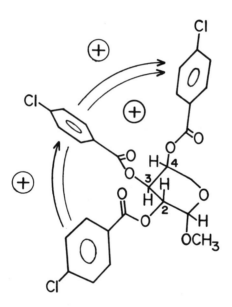

Figure 3-46. *Three positive exciton chiralities (represented by the three arrows) in the system of α -methyl-L-arabinopyranoside 2,3,4-tris(p-chlorobenzoate).*

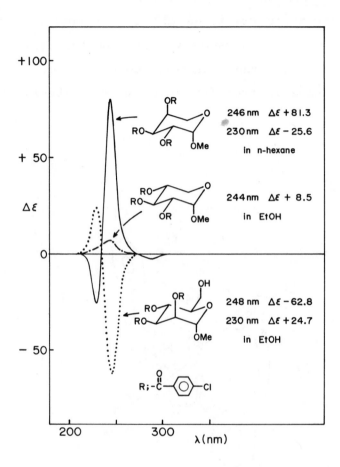

Figure 3-47. CD spectra of tris(p-chlorbenzoates) of sugars. (————) Methyl α-L-arabinopyranoside 2,3,4-tris(p-chlorobenzoate); (−•−•−) Methyl α-D-xylopyranoside 2,3,4-tris(p-chlorobenzoate); (••••••) Methyl α-D-mannopyranoside 2,3,4-tris(p-chlorobenzoate). [Adapted from references 12 and 64.]

Table 3-63.

+ (2-3)	245 (+78.1)	
	234 (0.0)	CD (64)
+ (3-4)	230 (-17.3)	
+ (2-4)	hexane	

131

− (2-3)	248 (-62.8)	
	236 (0.0)	CD (64)
− (3-4)	230 (+24.7)	
− (2-4)	EtOH	

132

− (2-3)	247 (-63.8)	
	238 (0.0)	CD (64)
− (3-4)	231 (+30.0)	
− (2-4)	EtOH	

133

Table 3-64.

	$\oplus$ (2-3)		
		244 (+8.3)	CD (64)
	$\ominus$ (3-4)		
		EtOH	
	O (2-4)		

134

In contrast, when opposing chiralities are present in a molecule, the
circular dichroic power of exciton interacting chromophores cancel out and no
clearcut Davydov split Cotton effects are observed.   The sugar tribenzoate
134, methyl α-D-xylopyranoside 2,3,4-tris(p-chlorobenzoate) belongs to this
category; the chiralities of the two vicinal moieties cancel out, while the
1,3-dibenzoate moiety is diequatorial.   Thus, the total exciton chirality is
zero, indicating no split Cotton effects.   In fact, compound **134** only shows
a weakly positive Cotton effect, λ_{ext} 244 nm, $\Delta\varepsilon$ +8.3, instead of the split CD
(Figure 3-47).

Table 3-65.

	$\ominus$ (3-4)	*237 (-68)*	
			CD (76)
	$\ominus$ (4-5)	*222 (+10)*	
	$\ominus$ (3-5)	CH_3CN	

135

The absolute configuration of an unusual aminoalditol isolated from a legume was established by application of the exciton chirality method.[76] Namely, blockade of the annular NH and freely rotating primary-OH was achieved by formation of the cyclic carbamate prepared from 1 mg of the alditol; the carbamate was subsequently benzoylated to yield derivative <u>135</u>.   Since the relative configurations of the three hydroxyls were determined by ^{1}H NMR to be 3-ax,4-eq,5-eq, they are spatially arranged to make exciton chiralities of the same sign; i.e., plus/plus/plus or minus/minus/minus.   The observed CD of carbamate benzoate <u>135</u> exhibited strong negative (-68) first and positive (+10) second Cotton effects at 237 nm and 222 nm, respectively; therefore, the absolute configuration with negative chiralities was assigned to the aminoalditol.

3-5-B.   Steroidal Tribenzoates

The present method is also applicable in exactly the same manner to more rigid steroid systems.

Table 3-66.

(+) (1-2)

(+) (2-3)

(+) (1-3)

247 (+85.6)
236 (0.0)   CD (64)
230 (-36.6)

X ray (3)

EtOH

136

The steroidal tribenzoate <u>136</u> of kogagenin, a 1β,2β,3α-trihydroxy steroidal saponin, exhibits very strong CD Cotton effects of the split type, λ_{ext} 247 nm, Δε +85.6 and 230 nm, Δε −36.6, in agreement with the three positive exciton chiralities (Figure 3-48).[64]

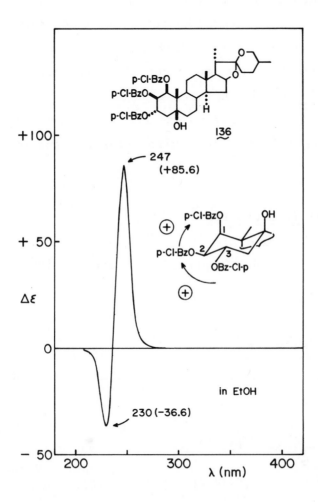

Figure 3-48. CD spectrum of kogagenin 1,2,3-tris(*p*-chlorobenzoate).

Table 3-67.

	280.2 (2,000)	281.0 (+0.6)
	273.0 (2,600)	274.0 (+0.7)
(+) (3-4)		253.5 (−1.0)
		235.8 (+23.1)
(0) (4-6)	*227.8 (35,000)*	228.0 (0.0) CD (5)
		222.0 (−19.6)
		X ray (3)
(+) (3-6)	0.9% dioxane/	10% dioxane/
	EtOH	EtOH

137

		282.2 (+0.9)
		274.8 (+1.4)
(+) (3-4)		*246.6 (+30.3)*
	237.2 (54,400)	238.8 (0.0) CD (5)
(0) (4-6)		*232.0 (−25.8)*
		X ray (3)
(+) (3-6)	0.9% dioxane/	10% dioxane/
	EtOH	EtOH

138

In the case of tribenzoates 137 and 138, the 1,2-dibenzoate moiety at 3β,4α positions and 1,4-dibenzoate moiety at 3β,6α positions are both positively twisted.[5] On the other hand, the exciton chirality of 1,3-dibenzoate moiety at 4α,6α positions is nil, because the two benzoate groups are aligned parallel to each other. Therefore, compounds 137 and 138 have two positive and one zero exciton chiralities which lead to the prediction of positive first and negative second Cotton effects. The observed CD spectra are consistent with this prediction, and the amplitudes are larger than those of the corresponding dibenzoates.

Table 3-68.

⊖ (2-3)	
O (2-11)	
⊖ (3-11)	

139

237 (-24.4)   CD (77)

220 (+20.0)   X ray (3)

EtOH

The interaction still operates between more remotely located tribenzoate systems.[77]   For example, metagenin tribenzoate (139) has three benzoate groups at 2β,3β, and 11α positions (A/B cis); the 2β,3β- and 3β,11α-dibenzoate moieties are left-handed, whereas the 2β and 11α benzoate groups are almost parallel to each other (see Figure 3-49).   Therefore, the three benzoate groups are located so as to generate Cotton effects of negative chirality. The tribenzoate 139 in fact exhibits negative first and positive second Cotton effects, λ_{ext} 237 nm, $\Delta\varepsilon$ -24.4 and 220 nm, $\Delta\varepsilon$ +20.   These $\Delta\varepsilon$ values are larger than the typical values for vicinal dibenzoates ($\Delta\varepsilon = 10 \sim 15$).

The same situation holds for the tribenzoate 140 of the phytoecdysone ajugasterone C,[77] which shows very intense split Cotton effects, λ_{ext} 237 nm, $\Delta\varepsilon$ -30 and 221 nm, $\Delta\varepsilon$ +34.   It is clear from Figure 3-49 that the Cotton effects associated with the ring B enone chromophore are too weak to interfere with the observation of the split Cotton effects.   These data confirm the C-2, -3, -11, and -5 (A/B cis) configuration in ajugasterone C and also indicate that ring A adopts the chair conformation.

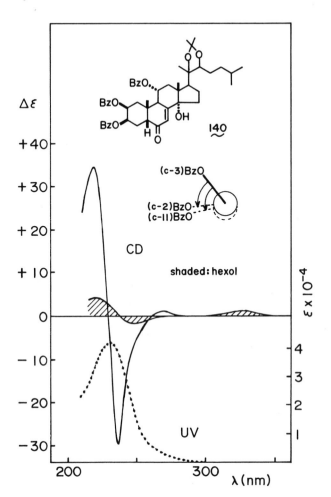

Figure 3-49. CD and UV spectra of ajugasterone C 20,22-acetonide 2,3,4-tri-
benzoate in ethanol: (••••••), UV, λ_{max} 231 nm (ε 40,300);   (———),   CD,
λ_{ext} 237 nm ($\Delta\varepsilon$ -30) and 221 nm ($\Delta\varepsilon$ +34).   The shaded CD spectrum is of
ajugasterone C itself: λ_{ext} 328 nm ($\Delta\varepsilon$ +1.5) n→π*,   and 250 nm ($\Delta\varepsilon$ -3.0) π→π*
of the enone system. [Adapted from reference 77.]

Table 3-69.

⊖ (2-3)	*237 (-30)*
○ (2-11)	*221 (+34)*
⊖ (3-11)	EtOH

CD (77)

X ray (22)

140

3-6.   An Additivity Correlation in the A Values of Interacting Chromophores

An unexpected additivity correlation has recently been found[78] in the amplitudes, or "A" values, of interacting three or more chromophores. Thus if I, II, and III are the three interacting chromophores, then the A value of the interacting I/II/III chromophore system can be approximated by the summation of component A values, A(I/II), A(II/III), and A(I/III) (Figure 3-50).

This observation was made in hexopyranose p-bromobenzoates as follows. Namely, over 40 hexopyranose bis, tris, and tetrakis(p-bromobenzoates) were prepared. Since the absorption maxima are at 244 nm, the polybenzoates exhibit split CD curves centered at 244 nm with either positive or negative "A" values depending on the spatial arrangement of the benzoate groups. Figures 3-51 through 3-54 list the found A values of various sugar benzoate α-methylpyranosides. Only the p-bromobenzoate groups are denoted by shaded circles;

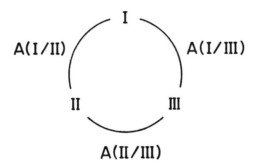

$$A \approx A(I/II) + A(II/III) + A(I/III)$$

Figure 3-50. *The A value of an interacting triple system can be approximated by the summation of component A values.*

the unsubstituted carbons 2, 3, 4, and 5 being any other nonbenzoylated group such as CH_2 (desoxysugar, e.g., xylose), CHOH (free hydroxyl), CHOAc (acetate), CHOMe (methyl ether), etc. As mentioned above, the C-1 substituent was an α-OMe in most cases.

When two or more pyranose benzoates belonging to a particular class, i.e., class I-V, were measured, it was found that the A values agreed within an error of ± 3. Figure 3-51 lists the A values of various 1,2-vicinal dibenzoates. Class I lists the experimental A values of nine various 2e,3e-dibenzoates (exciton coupling unit 1) and two each of 3e,4e-(unit 2), 3e,4a-(unit 3), and 2a,3e-dibenzoates (unit 4). It was found that, disregarding the signs which of course reflect the chiralities, the A values for these 1,2ee and 1,2ea vicinal dibenzoate units all fall within the range of 62±3. The two examples of 1,2aa dibenzoates (unit 5) had values of 6±3. If the two dibenzoates had been truly diaxial, no split CD should have been observed. The observation of the small split CD indicates that the two benzoate groups are

"A" Values of p-Br-Benzoates in MeOH

I) di-OBz, 1,2ee and ea: A = 62

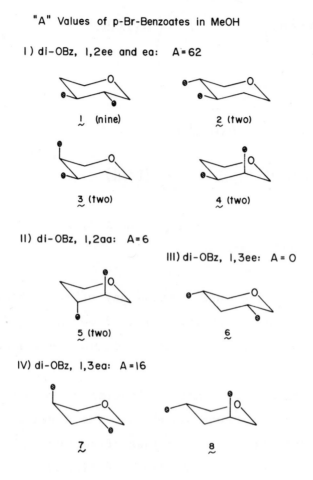

$\underset{\sim}{1}$ (nine) $\underset{\sim}{2}$ (two)

$\underset{\sim}{3}$ (two) $\underset{\sim}{4}$ (two)

II) di-OBz, 1,2aa: A = 6

III) di-OBz, 1,3ee: A = 0

$\underset{\sim}{5}$ (two) $\underset{\sim}{6}$

IV) di-OBz, 1,3ea: A = 16

$\underset{\sim}{7}$ $\underset{\sim}{8}$

Figure 3-51. The "A" values of various α-methylpyranoside   bis(p-bromobenzoates), where shaded circles denote p-bromobenzoate group; the unsubstituted carbons 2,3,4, and 5 can be any other nonbenzoylated groups such as CH₂, CHOH, CHOAc, CHOMe, etc. Numerals in parentheses indicate number of samples measured. For example, the A values of nine sugars with 2-equatorial and 3-equatorial benzoate groups (unit 1) fall within 62±3.

tilted from an ideal 180° <u>trans</u>-diaxial relation, probably due to 1,3–diaxial interactions (see section 3–1–C and Figure 3–13 for a steroidal case).

The A values for a 1,3<u>ee</u> dibenzoate (unit <u>6</u>) and two 1,3<u>ea</u> dibenzoates (units <u>7</u> and <u>8</u>) are also recorded.   In the coupled unit <u>6</u>, the two benzoate groups are in the same plane, hence the A value is zero.   In units <u>7</u> and <u>8</u>, the two benzoate groups are more distant than in the vicinal dibenzoates

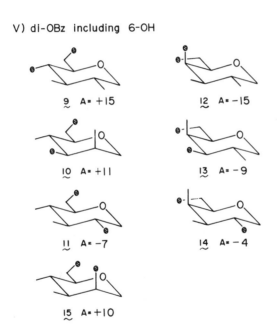

V) di-OBz including 6-OH

<u>9</u> A = +15   <u>12</u> A = -15

<u>10</u> A = +11   <u>13</u> A = -9

<u>11</u> A = -7   <u>14</u> A = -4

<u>15</u> A = +10

Figure 3-52.   The observed A values of sugar bis(p-bromobenzoates) including 6-hydroxyl group, where nonbenzoylated substituents are denoted by lines.   See text for conformation of the 6-benzoate group.

listed in class I, and thus the smaller A value of +16 is obtained. Recall that other factors being equal, the amplitude is inversely proportional to the square of the interchromophoric distance, as described in section 3-1-A; see also section 10-8 for theoretical derivation.

The A values of dibenzoates with the 6-benzoate group are recorded in Figure 3-52, class V; here the nonbenzoylated substituents are donated by lines. The signs of A values for exciton coupling units 9-11 and 15, which bear 4-equatorial substituents, show that the conformations of the 6-benzoate group must be as depicted; in contrast, when 4-axial substituents are present as in units 12-14 the 6-benzoate points to the "left." This effect of the 4-substituent on the conformation of the 6-benzoate can be visualized by space-filling molecular models.

By employing these A values for various exciton coupling units, we can now calculate the A values of tri- and tetra-benzoate systems. In the case of 1,2,3eee-tribenzoate system 16 in Figure 3-53, no split CD was observed. Namely, the chirality between 2,3- is positive (or +62), that between 3,4- is negative (or -62), and that between 2,4- is zero; they cancel out[64] and the value is zero. The values for 1,2,3eea-tribenzoate systems 17 and 18 are large because all of the chiralities of component coupling units are of the same sign[64]; in the case of system 17, A(calcd)= +62+62+16=+140, while the observed values were about +135.

Other examples of the "additivity" relationships for tribenzoates, including the 6-position, are given in Figure 3-53.

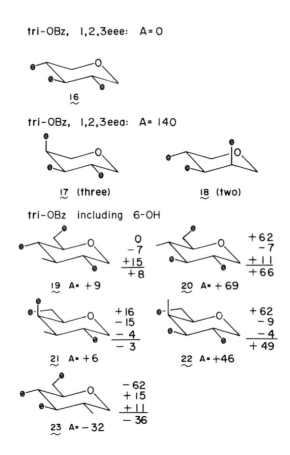

tri-OBz, 1,2,3eee: A = 0

16

tri-OBz, 1,2,3eea: A = 140

17 (three)   18 (two)

tri-OBz including 6-OH

$$\begin{array}{r} 0 \\ -7 \\ +15 \\ \hline +8 \end{array}$$

19   A = +9

$$\begin{array}{r} +62 \\ -7 \\ +11 \\ \hline +66 \end{array}$$

20   A = +69

$$\begin{array}{r} +16 \\ -15 \\ -4 \\ \hline -3 \end{array}$$

21   A = +6

$$\begin{array}{r} +62 \\ -9 \\ -4 \\ \hline +49 \end{array}$$

22   A = +46

$$\begin{array}{r} -62 \\ +15 \\ +11 \\ \hline -36 \end{array}$$

23   A = -32

Figure 3-53.   The observed and calculated A values of sugar tris(p-bromoben-zoates).

The additivity relation was found to hold even for the tetrabenzoate systems of α- and β-methyl-glucosides 24, α-methyl-galactoside 25, and α-methyl-mannoside 26 (the A values of component dibenzoate coupling units and calculated final A values are listed in Figure 3-54). The agreement found between the observed and calculated values, especially in view of the fact that six component values have to be considered for these three cases, is quite remarkable.

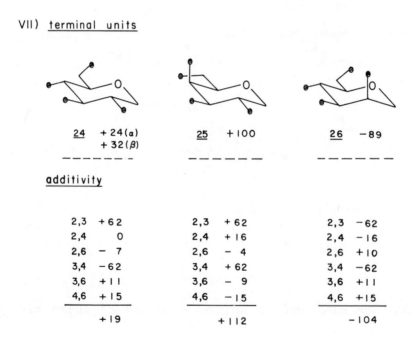

VII) <u>terminal units</u>

<u>24</u>	+24(α)	<u>25</u>	+100	<u>26</u>	−89
	+32(β)				

<u>additivity</u>

2,3	+62	2,3	+62	2,3	−62
2,4	0	2,4	+16	2,4	−16
2,6	−7	2,6	−4	2,6	+10
3,4	−62	3,4	+62	3,4	−62
3,6	+11	3,6	−9	3,6	+11
4,6	+15	4,6	−15	4,6	+15
	+19		+112		−104

Figure 3-54. *The observed and calculated A values of sugar tetrakis(p-bromo-benzoates).*

Although such an additivity correlation may not immediately be obvious from theory, the data mentioned with pyranose benzoates show that it is true. If exploited properly, this additivity relation should clearly expand the utility of the exciton chirality method, especially when dealing with complex multichromophore systems.[79]

References

1. S.-M. L. Chen, N. Harada, and K. Nakanishi, J. Am. Chem. Soc. 96, 7352 (1974).

2. N. Harada, S.-M. L. Chen, and K. Nakanishi, J. Am. Chem. Soc. 97, 5345 (1975).

3. Absolute configurations of steroidal compounds are established by the X-ray Bijvoet method; see W. Klyne and J. Buckingham, Atlas of Stereo-chemistry (London: Chapman and Hall, 1978) vol 1, pp. 121-26; vol. 2, pp. 63-64.

4. N. Harada and K. Nakanishi, J. Am. Chem. Soc. 91, 3989 (1969).

5. N. Harada, unpublished data.

6. R. U. Lemieux, Ann. N. Y. Acad. Sci. 222, 915 (1973).

7. R. F. Curl, J. Chem. Phys. 30, 1529 (1959).

8. A. McL. Mathieson, Tetrahedron Lett. 1965, p. 4137.

9. Syntex Analytical Instruments, unpublished data.

10. E. J. Gabe and W. H. Barnes, Acta Crystallogr. 16, 796 (1963).

11. N. Harada, S. Suzuki, H. Uda, and K. Nakanishi, J. Am. Chem. Soc. 93, 5577 (1971).

12. N. Harada and K. Nakanishi, <u>Acc. Chem. Res.</u> <u>5</u>, 257 (1972).

13. V. Delaroff, a private communication, 1974.

14. X-ray determination of (+)-endo-3-bromocamphor: M. G. Northolt and J. H. Palm, <u>Rec. Trav. Chim.</u> <u>85</u>, 143 (1966).

15. F. H. Allen and D. Rogers, <u>Chem. Commun.</u> 1966, p. 836.

16. J. M. Bijvoet, A. F. Peerdeman, and A. J. Van Bommel, <u>Nature</u> <u>168</u>, 271 (1951).

17. I. Hanazaki and H. Akimoto, <u>J. Am. Chem. Soc.</u> <u>94</u>, 4102 (1972).

18. S. F. Mason, R. H. Seal, and D. R. Roberts, <u>Tetrahedron</u> <u>30</u>, 1671 (1974).

19. G. Snatzke, in <u>Optical Activity and Chiral Discrimination</u>, ed. S. F. Mason, Dordrecht: D. Reidel, 1979. Chapter 2, page 41.

20. American Cyanamid Co. French Patent 1 482 866, 1967; <u>Chem. Abstr.</u> <u>69</u>, 19143v (1968).

21. J. Decombe, <u>Bull. Soc. Chim. Fr.</u> 1951, p. 416.

22. Absolute configurations of ecdysones have been determined by various physical data including relative X-ray data and chemical syntheses from known steroids; see K. Nakanishi, T. Goto, S. Ito, S. Natori, and S. Nozoe, eds., <u>Natural Products Chemistry</u> (Tokyo: Kodansha; and New York: Academic Press, 1974) vol. 1, p. 525.

23.   M. Koreeda and K. Nakanishi, J. Chem. Soc., Chem. Commun. 1970, p. 351.

24.   Synthesis from (-)-α-santonin: A. Murai, K. Nishizakura, N. Katsui, and
      T. Masamune, Tetrahedron Lett. 1975, p. 4399.   A. Murai, K. Nishizakura,
      N. Katsui, and T. Masamune, Bull. Chem. Soc. Jpn. 50, 1206 (1977).

25.   N. Katsui, H. Kitahara, F. Yagihashi, A. Matsunaga, and T. Masamune,
      Chem. Lett. 1976, p. 861.   N. Katsui, A. Matsunaga, H. Kitahara, F.
      Yagihashi, A. Murai, T. Masamune, and N. Sato, Bull. Chem. Soc. Jpn. 50,
      1217 (1977).

26.   G. I. Birnbaum, C. P. Huber, M. L. Post, J. B. Stothers, J. R. Robinson,
      A. Stoessl, and E. W. B. Word, J. Chem. Soc., Chem. Commun. 1976, p.
      330.

27.   M. Shiro, T. Sato, H. Koyama, Y. Maki, K. Nakanishi, and S. Uyeo, Chem.
      Commun. 1966, p. 98.

28.   M. Shiro and H. Koyama, J. Chem. Soc. B 1971, p. 1342.

29.   T. Ito, N. Harada, and K. Nakanishi, Agr. Biol. Chem. (Japan) 35, 797
      (1971).

30.   Synthesis: J. E. Hay and L. J. Haynes, J. Chem. Soc. 1958, p. 2231.

31.   N. Harada and K. Nakanishi, J. Chem. Soc., Chem. Commun. 1970, p. 310.

32.   A. Furusaki, H. Shirahama, and T. Matsumoto, Chem. Lett. 1973, p. 1293.

33. M. Kuroyanagi, K. Yoshihira, and S. Natori, Chem. Pharm. Bull. 19, 2314 (1971).

34. Y. Yamada, C.-S. Hsu, K. Iguchi, M. Suzuki, H.-Y. Hsu, and Y.-P. Chen, Tetrahedron Lett. 1974, p. 2513.

35. H. Bernotat-Wulf, A. Niggli, L. Ulrich, and H. Schmid, Helv. Chim. Acta 52, 1165 (1969).

36. H. Ziffer, D. M. Jerina, D. T. Gibson, and V. M. Kobal, J. Am. Chem. Soc. 95, 4048 (1973).

37. V. M. Kobal, D. T. Gibson, R. E. Davis, and A. Garza, J. Am. Chem. Soc. 95, 4420 (1973).

38. J. I. Seeman and H. Ziffer, J. Org. Chem. 39, 2444 (1974).

39. Y. Tsuda and T. Fujimoto, J. Chem. Soc., Chem. Commun. 1970, p. 260.

40. X-ray Bijvoet determination of 21-oxo-serrat-13-en-3β-ol: F. H. Allen and J. Trotter, Acta Crystallogr. suppl. 1969, S137.

41. J. Sakakibara, K. Ikai, and M. Yasue, Chem. Pharm. Bull. 20, 861 (1972).

42. X-ray Bijvoet determination of grayanotoxin I: D. W. Engel, K. Zechmeister, and W. Hoppe, Tetrahedron Lett. 1972, p. 1323.

43. C. W. Lyons and D. R. Taylor, J. Chem. Soc., Chem. Commun. 1976, p. 647.

44. A. Glosse and H.-P. Sigg, Helv. Chim. Acta 56, 619 (1973).

45. M. Nukina and S. Marumo, Tetrahedron Lett. 1977, p. 3271.

46. T. Sassa, A. Takahama, and T. Shindo, Agric. Biol. Chem. 39, 1729 (1975).

47. T. Yoshida, J. Nobuhara, M. Uchida, and T. Okuda, Tetrahedron Lett. 1976, p. 3717.

48. H. Kaise, M. Shinohara, W. Miyazaki, T. Izawa, Y. Nakano, M. Sugawara, K. Sugiura, and K. Sasaki, J. Chem. Soc., Chem. Commun. 1979, p. 726.

49. K. Nakanishi, H. Kasai, H. Cho, R. G. Harvey, A. M. Jeffrey, K. W. Jennette, and I. B. Weinstein, J. Am. Chem. Soc. 99, 258 (1977).

50. H. Yagi, H. Akagi, D. R. Thakker, H. D. Mah, M. Koreeda, and D. M. Jerina, J. Am. Chem. Soc. 99, 2358 (1977).

51. D. H. R. Barton, H. T. Cheung, A. D. Cross, L. M. Jackman, and M. Martin-Smith, J. Chem. Soc. 1961, p. 5061.

52. I. C. Paul, G. A. Sim, T. A. Hamor, and J. M. Robertson, J. Chem. Soc. 1962, p. 4133.

53. S. Hosozawa, N. Kato, and K. Munakata, Tetrahedron Lett. 1974, p. 3753.

54. N. Harada and H. Uda, J. Am. Chem. Soc. 100, 8022 (1978).

55. N. Kato, S. Shibayama, K. Munakata, and C. Katayama, <u>Chem. Commun.</u> 1971, p. 1632.

56. N. Kato, K. Munakata, and C. Katayama, <u>J. Chem. Soc., Perkin Trans. 2</u> 1973, p. 69.

57. N. Kato, M. Shibayama, and K. Munakata, <u>J. Chem. Soc., Perkin Trans. 1</u> 1973, p. 712.

58. D. Rogers, G. G. Unal, D. J. Williams, S. V. Ley, G. A. Sim, B. S. Joshi, and K. R. Ravindranath, <u>J. Chem. Soc., Chem. Commun.</u> 1979, p. 97.

59. G. Trivedi, H. Komura, I. Kubo, K. Nakanishi, and B. S. Joshi, <u>J. Chem. Soc., Chem. Commun.</u> 1979, p. 885.

60. J. MacMillan and T. J. Simpson, <u>J. Chem. Soc., Perkin Trans. 1</u> 1973, p. 1487.

61. W. J. McGahren, G. A. Ellestad, G. O. Morton, M. P. Kunstzmann, and P. Mullen, <u>J. Org. Chem.</u> <u>38</u>, 3542 (1973).

62. Asymmetric synthesis of natural pestalotin: D. Seebach and H. Meyer, <u>Angew. Chem. Internat. Edit.</u> <u>13</u>, 77 (1974).

63. Synthesis of enantiomeric pestalotin from (R,R)-(+)-tartaric acid: H. Meyer and D. Seebach, <u>Justus Liebigs Ann. Chem.</u> 1975, p. 2261. From (R)-(+)-glyceraldehyde: K. Mori, M. Oda, and M. Matsui, <u>Tetrahedron Lett.</u> 1976, p. 3173.

64. N. Harada, H. Sato, and K. Nakanishi, J. Chem. Soc., Chem. Commun. 1970, p. 1691.

65. N. Kato and Y. Kojima, "Abstracts of Papers," 8th Symposium on the Structural Organic Chemistry, Kyoto, Japan, 1975, p. 56.

66. H. Gerlach, Helv. Chim. Acta 51, 1587 (1968).

67. N. Harada, N. Ochiai, K. Takada, and H. Uda, J. Chem. Soc., Chem. Commun. 1977, p. 495.

68. R. S. Cahn, C. Ingold, and V. Prelog, Angew. Chem. Internat. Edit. 5, 385 (1966).

69. M. Tichy, Tetrahedron Lett. 1972, p. 2001.

70. R. Miura, J. Okada, and M. Nakazaki, "Abstracts of Papers," 15th Symposium on the Chemistry of Natural Products, Nagoya, Japan, 1971, p. 69.

71. H. Chikamatsu, H. Murakami, and M. Nakazaki, "Abstr. No 2U40," National Meeting of the Chemical Society of Japan, Higashiosaka, April 1977.

72. A. W. Johnson, R. M. Smith, and R. D. Guthrie, J. Chem. Soc., Perkin Trans. 1, 1972, p. 2153.

73. M. Kawai, U. Nagai, and M. Katsumi, Tetrahedron Lett. 1975, p. 3165.

74. U. Nagai, private communication, 1976.

75. Y. Saito and H. Iwasaki, Bull. Chem. Soc. Jpn. 35, 1131 (1962).

76. L. E. Fellows, E. A. Bell, D. G. Lynn, F. Pilkiewicz, I. Miura, and K. Nakanishi, J. Chem. Soc., Chem. Commun. 1979, p. 977.

77. M. Koreeda, N. Harada, and K. Nakanishi, J. Chem. Soc., Chem. Commun. 1969, p. 548.

78. H. -W. Liu and K. Nakanishi, J. Am. Chem. Soc. 103, 5591 (1981); H. -W. Liu and K. Nakanishi, J. Am. Chem. Soc., 104, 1178 (1982).

79. Application to oliogosaccharides: H. -W. Liu and K. Nakanishi, J. Am. Chem. Soc., 103, 7005 (1981).

80. Angular dependence of UV λ_{max} in exciton coupling systems: N. Harada and H. Uda, J. Chem. Soc., Chem. Commun. 1982, p. 230.

81. (a) Absolute stereochemistry of 2,2'-spirobi-indane-1,1'-diols and -1,1' -dione: N. Harada, T. Ai, and H. Uda, J. Chem. Soc., Chem. Commun. 1982, p. 232. (b) A. Meyer, H. Neudeck, and K. Schlögl, Tetrahedron Lett. 1976, p. 2233.

82. Absolute configuration of brevetoxin B, a red tide toxin: Y. Y. Lin, M. Risk, S. M. Ray, D. V. Engen, J. Clardy, J. Golik, J. C. James, and K. Nakanishi, J. Am. Chem. Soc. 103, 6773 (1981).

IV. APPLICATION OF THE EXCITON CHIRALITY METHOD TO POLYACENE AND RELATED CHROMOPHORES

4-1. Circular Dichroism and Absolute Configuration of Ethanodibenz[a,h]anthracene and Its Derivatives[1-3]

As discussed in the case of 6,15-dihydro-6,15-ethanonaphtho[2,3-c]penta-phene[1,3,4] in section 1-4, the exciton chirality method is applicable to the intense ^1B_b transition of polyacene chromophoric systems, which is polarized along the long axis of the chromophores.

Table 4-1.

	325.0 (2,500)	326.0 (−4.3)	
		322.0 (+5.0)	
		304.5 (+25.5)	
	283.5 (11,100)	283.0 (−35.6)	CD (2),(3)
		237.0 (+326.5)	
	232.5 (98,200)	229.5 (0.0)	Correl. (2)
		223.5 (−180.5)	
		205.0 (−115.0)	
1	EtOH	EtOH	

190

In the case of $(7\underline{R},14\underline{R})$-$(+)$-7,14-dihydro-7,14-ethanodibenz[$\underline{a},\underline{h}$]anthracene ($\underline{1}$), which has two naphthalene chromophores in a chiral relation, the CD spectrum shows typical and very strong split Cotton effects: λ_{ext} 237 nm, $\Delta\varepsilon$ +326.5 and λ_{ext} 224 nm, $\Delta\varepsilon$ -180.5; A $(= \Delta\varepsilon_1 - \Delta\varepsilon_2) = +507.0$, at the ^1B_b transition (λ_{max} 232.5 nm, ε 98,200) (Figure 4-1). Namely, the positive sign of the A value is in agreement with the positive chirality between the two

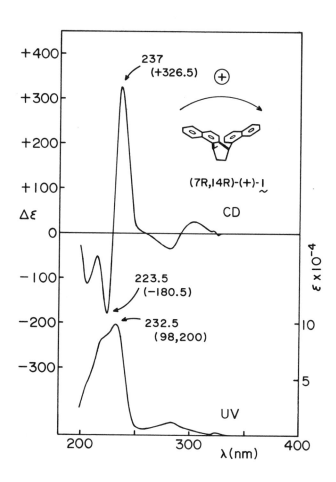

Figure 4-1. *CD and UV spectra of $(7\underline{R},14\underline{R})$-$(+)$-7,14-dihydro-7,14-ethanodibenz [$\underline{a},\underline{h}$]anthracene in ethanol.* [Adapted from reference 2.]

long axes of naphthalene chromophores. Thus, the (7R,14R) absolute configuration of (+)-1 is established.[2] The absolute stereochemistry has also been determined by chemical correlations and X-ray analyses.

Table 4-2.

2	(+)	326.3 (1,000) 286.5 (11,700) *236.8 (98,100)* 226.0 (95,300)	326.5 (-2.3) 307.0 (+25.1) 281.0 (-18.7) *240.6 (+371.5)* *232.6 (0.0)* *227.0 (-149.4)*	CD (1),(2) Correl. (2)
		0.18% dioxane/ EtOH	0.18% dioxane/ EtOH	

3	(+)	293.0 (16,000) *238.5 (116,200)*	312.0 (+27.5) 283.0 (-27.1) *242.0 (+340.3)* *230.5 (0.0)* *225.0 (-87.6)* 212.5 (-139.3)	CD (1),(2) Correl. (2)
		EtOH	EtOH	

The same situation is true in the cases of dimethyl (2) and tetramethyl (3) derivatives.[1,2]

Table 4-3.

	324 (2,100)	330 (+0.3)
	315 (1,900)	326 (-0.1)
	283 (12,200)	304 (+28.6)
	232 (95,200)	282 (-41.8)
	221 (81,700)	*236 (+306)*
		222 (-195)
		206 (-126)

4

CH$_3$CN   CH$_3$CN   CD (5)

Hagishita, et al.,[5] have reported the CD spectrum of a similar compound (4) for determining the absolute configuration from the chiral exciton coupling. The derived absolute stereochemistry was in agreement with that deduced by a kinetic resolution method.

4-2. 1,1'-Bianthryl and 1,1'-Binaphthyl Derivatives and Related Compounds

The following examples of 1,1'-bianthryl and 1,1'-binaphthyl derivatives are very interesting and quite significant in the development of the CD exciton coupling method. Namely, Mason and his coworkers[6-8] have first reported that the absolute configurations of 1,1'-bianthryl derivatives are unambiguously determinable by means of CD spectroscopy in a nonempirical manner.

   Dimethyl 1,1'-bianthryl-2,2'-dicarboxylate (-)-(5) exhibits very intense negative first and positive second Cotton effects at the ^1B_b transition of the anthracene chromophore: λ_{ext} 267 nm, $\Delta\varepsilon$ -760 and 250 nm, $\Delta\varepsilon$ +780.

Table 4-4.

		404 (7,900)	403 (+13)	
		383 (10,000)		CD (7),(8)
		364 (7,900)		
		343 (6,300)		UV (9)
		268 (125,900)	*267 (-760)*	
			250 (+780)	Correl.
				(12),(13)
		cyclohexane/	cyclohexane	
(R)-(-)-5		dioxane		X ray (11)

It is thus concluded that the compound has a negative chirality between the two long axes of anthracene moieties; i.e., the absolute configuration is R.

It should be noted that in the present case, the assignment of (R) configuration is independent of the rotational conformation around the chiral axis connecting the two anthracene groups.[7,8] In other words, the sign of the first Cotton effect is always negative when the dihedral angle between two anthracene planes is changed from 0° through 180°.

Table 4-5.

		422 (15,800)	420 (+15)	
		400 (12,600)		
			267 (-1100)	CD (7)
		248 (199,500)	*248 (+1100)*	
				UV (9)
		cyclohexane/	cyclohexane/	
(R)-6		dioxane	dioxane	

In the case of compound **6**, since the dihedral angle is fixed at ca. 30° by an ethano bridge, the absolute configuration is directly assignable from the observed CD spectrum; the enantiomer with (**R**) absolute configuration exhibits negative first and positive second Cotton effects at the ^1B_b transition region.

The (**R**) and (**S**) nomenclature of biphenyl, binaphthyl, and bianthryl compounds with axial chirality is defined as follows:[14] the structure is regarded as an elongated tetrahedron and either one of the two pairs of ortho carbons is assigned the first two priorities as shown in Figure 4-2. In the case of chiral 6-methoxy-[1,1'-biphenyl]-2,2'-diamine (**7**) depicted in Figure 4-2, the sequence rule path is anticlockwise (**S**).

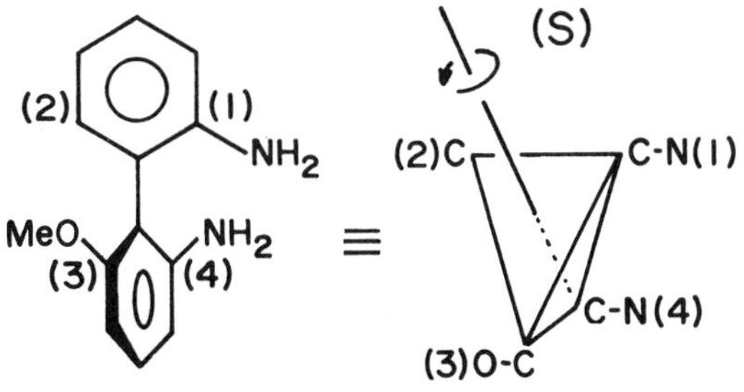

7

Figure 4-2. *(S) configuration of 6-methoxy-[1,1'-biphenyl]-2,2'-diamine.*

The determination of the absolute stereochemistries of these very interesting and important compounds by the X-ray crystallographic Bijvoet method was carried out by Akimoto et al.;[11] they determined the absolute configuration of compound $\underline{8}$ by using the heavy atom effect of bromobenzene which is contained in the crystal as crystal solvent. Various binaphthyl and bianthryl compounds were then correlated with this key compound.[10,12,13] Since compounds $\underline{5}$ and $\underline{6}$ have also been chemically correlated, the absolute stereochemistries as determined by the CD chiral exciton coupling method are confirmed by the X-ray method.

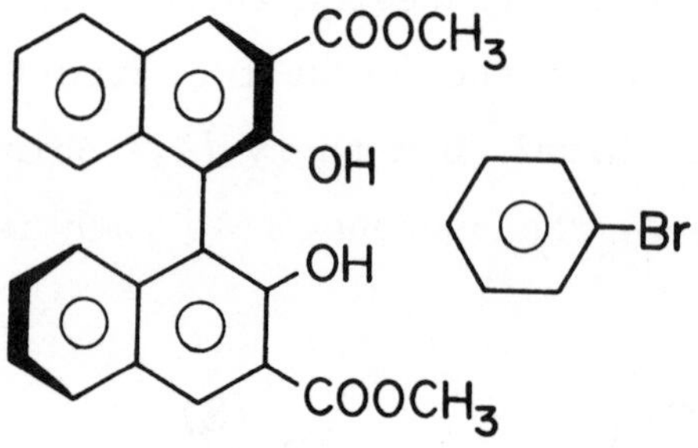

(R)-(+)-$\underline{8}$

The absolute configuration of (−)-2,2'-bis(bromomethyl)-1,1'-binaphthyl ($\underline{9}$) was also determined to be $\underline{S}$ by the X-ray Bijvoet method.[15]

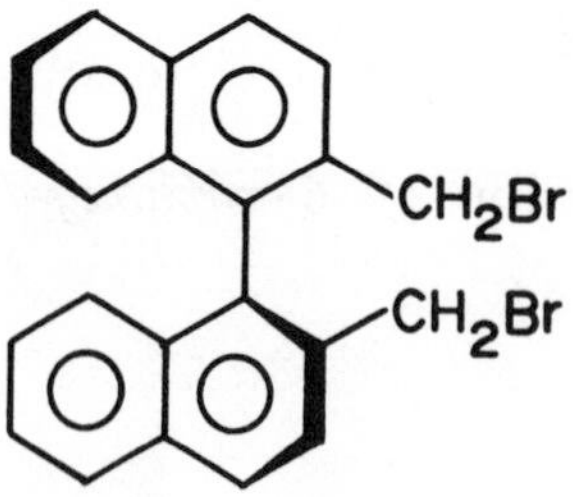

(S)-(−)-$\underline{9}$

In the case of chiral 1,1'-binaphthyl compounds, the ^1B_b transition that exhibits exciton split Cotton effect is around 225 nm. For instance, (S)-1,1'-binaphthyl-2,2'-dimethanol (10) shows very intense positive first and negative second Cotton effects at the ^1B_b transition band: λ_{ext} 231.3 nm, $\Delta\varepsilon$ +342.4 and λ_{ext} 224.3 nm, $\Delta\varepsilon$ -329.0 (Figure 4-3).[16] On the other hand, the ^1L_a transition around 280 nm exhibits only a single Cotton effect of weak intensity.

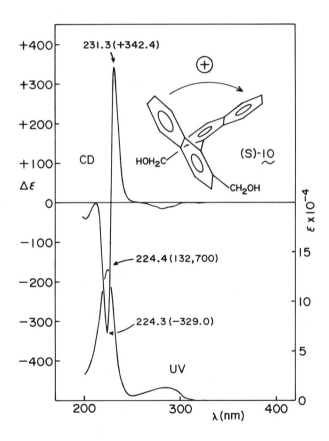

Figure 4-3. CD and UV spectra of (S)-1,1'-binaphthyl-2,2'-dimethanol in ethanol.

Table 4-6.

	320.2 (800)	320.6 (−0.5)		
	283.3 (13,600)	283.5 (−12.7)		
		231.3 (+342.4)	CD (16)	
	224.4 (132,700)	228.0 (0.0)		
		224.3 (−329.0)	Correl.	
		212.8 (+4.6)	(10),(13)	

(S)-10   EtOH   EtOH

In the present case, the positive sign of the first Cotton effect is correlated with a right-handed screwness between the two long axes of naphthalene chromophores. The chiral exciton coupling method is therefore applicable to this system.

In a similar way, the absolute stereochemistry of a number of binaphthyl compounds has been correlated with typical exciton split Cotton effects of the ^1B_b transition; enantiomers having a right-handed screwness, in general, exhibit positive first and negative second Cotton effects at the ^1B_b transition, as exemplified by the binaphthyl compounds listed in Table 4-7.

Table 4-7.

	293.5 (14,000)			
	283 (14,300)	280 (−2.4)		
		260 (+1.1)	CD (8),(17),	
		225 (+250)	(18)	
	220 (108,000)	220 (0)		
		214 (−179)		
			Correl.	
			(10),(13)	

(S)-11   96% EtOH   95% EtOH

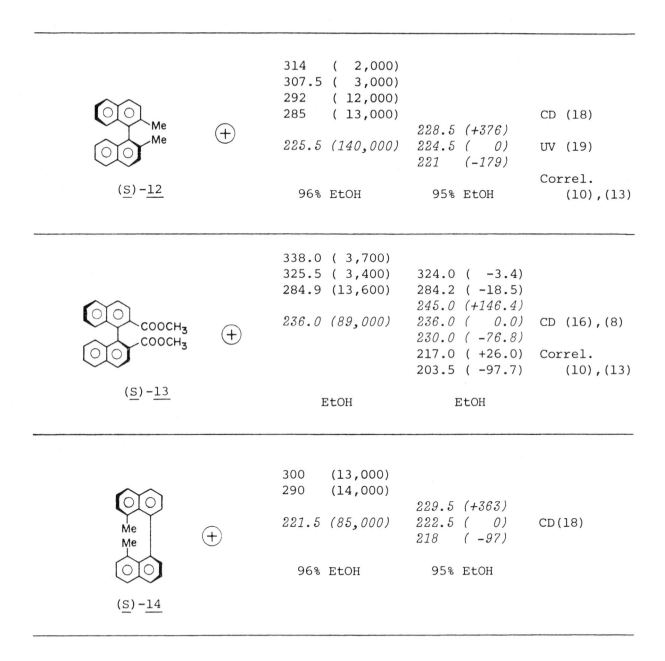

314   (2,000)
307.5 (3,000)
292   (12,000)
285   (13,000)   CD (18)
                             228.5 (+376)
225.5 (140,000)   224.5 (0)   UV (19)
                             221   (-179)
                                               Correl.
(S)-12   96% EtOH   95% EtOH   (10),(13)

338.0 (3,700)
325.5 (3,400)   324.0 (-3.4)
284.9 (13,600)   284.2 (-18.5)
                             245.0 (+146.4)
236.0 (89,000)   236.0 (0.0)   CD (16),(8)
                             230.0 (-76.8)
                             217.0 (+26.0)   Correl.
                             203.5 (-97.7)   (10),(13)

(S)-13   EtOH   EtOH

300   (13,000)
290   (14,000)
                             229.5 (+363)
221.5 (85,000)   222.5 (0)   CD(18)
                             218   (-97)

(S)-14   96% EtOH   95% EtOH

	332	(9,400)		
	309	(13,800)		
	239.5	*(44,000)*	*240*	*(+127)*	CD (8),(18)
			232.5	*(0)*	
	219.5	(60,000)	*223*	*(-151)*	
		96% EtOH		95% EtOH	

(S)-15

			371.0	(-0.6)	
	347.5	(6,300)	337.6	(+12.9)	
	282.0	(13,300)	293.0	(+18.2)	
			248.4	*(-221.4)*	CD (8),(8a)
	239.6	*(95,500)*	*239.8*	*(0.0)*	
			233.0	*(+117.2)*	Correl.
	213.0	(42,800)	217.0	(-29.8)	(10),(13)
			207.0	(+89.1)	
		EtOH		EtOH	

(R)-(+)-16

The present assignment has been established by the X-ray and chemical correlation studies.[10,13]

It should be noted that the (R) and (S) nomenclature is not directly correlated to the screwness between the two long axes of naphthalene chromophores. In the case of C_2-symmetrically substituted binaphthyl compounds, however, the (S) configuration corresponds to a clockwise screwness. Therefore, it is concluded that the (S)-enantiomer of C_2-symmetrically substituted binaphthyls would exhibit positive first and negative second Cotton effects.

From the theoretical viewpoint, the situation is more complex; theoretical calculations obtained by Hanazaki, et al.[20] and by Mason, et al.,[8] indicate that the sign of exciton split CD Cotton effects of 1,1'-binaphthyls changes from plus to minus when the dihedral angle of the (S)-enantiomer with right-handed screwness is changed from 0° to 180°; the zero point being around 110°. Therefore, absolute configuration of binaphthyl compounds is not independently determinable from the CD data alone. Namely, information on the dihedral angle is needed. However, since the chemical correlation studies by Akimoto and coworkers[10,13] have established the absolute configurations of these compounds, it is now inversely concluded that the binaphthyl derivatives 10–16 adopt conformations in which the dihedral angle is less than 110°. In fact, X-ray crystallographic data of some binaphthyl compounds reveal that the dihedral angle is distributed in the range of 68°–92°.[15]

4-3. Application to Natural Products with Binaphthyl Chromophore

Applications of the above results to natural products are as follows.

Table 4-8.

	336 (14,500)		
	322 (14,500)		
	308 (14,500)		
	245sh (52,500)	_245 (-91.0)_	
	232 (83,200)	_235 (0.0)_	CD (21)
		_225 (+25.1)_	
	EtOH	EtOH	

17

For example, the isoquinoline alkaloid derivative <u>17</u> with naphthalene and isoquinoline chromophores exhibits negative first and positive second Cotton effects at the 1_{B_b} transition region.   Therefore, the absolute configuration with left-handed screwness was assigned to the alkaloid.[21]

Table 4-9.

<u>18</u>	+	*243 (+95.0)* *228 (-86.4)* EtOH CD (22)

In the case of derivative <u>18</u>, although the naphthalene chromophore is attached to the isoquinoline chromophore at the β–position, theoretical consideration indicates that the results discussed above still hold for this case. Therefore, since the first Cotton effect is positive, the absolute stereochemistry of right-handed screwness is assigned to <u>18</u>.[22]

Table 4-10.

<u>19</u>	+	417 (27,900) 334 (17,100) 320 (21,800) *278 (114,000)* 233 (70,100) MeOH	*273 (+89.4)* *235 (-146.0)* MeOH CD (23)

413 (21,400)
331 (11,000)
317 (14,200)

285 (+140.6)

278 (86,700)   CD (23)

269 (-193.3)

231 (47,900)

MeOH   MeOH

20

In a similar way, Endo, et al.,[23] determined the absolute configuration of antimicrobial and antispasmodic dimeric tetrahydroanthracenes, singueanol-I (19) and II (20), isolated from Cassia singueana, an East African medi-
cinal plant. Both dimers have two naphthalene chromophores which interact
with each other to give exciton split Cotton effects of positive chirality.
Therefore, absolute configurations with positive chirality were assigned. In
singueanol-I 19, however, the first Cotton effect is shifted to shorter
wavelengths than the UV maximum. Blocking of the phenolic groups by methylation and thus eliminating the effect of strong chelations probably would have
led to more typical split CD data. On the other hand, singueanol-II 20 shows
typical exciton split Cotton effects at the ^1B_b transition.

4-4. Absolute Stereochemistry and Chiroptical Properties of Chiral Triptycenes Composed of Three Aromatic Chromophores

The following triptycene derivatives provide examples of exciton split CD
Cotton effects due to chiral exciton coupling between three aromatic chromophores.[24,25]

The CD spectrum of (5$\underline{R}$,12$\underline{R}$)-(-)-1,15-diethynyl-5,12-dihydro-5,12[1',2'] benzenonaphthacene (<u>21</u>) exhibits intense negative first and positive second Cotton effects, λ_{ext} 245.5 nm, $\Delta\varepsilon$ -138.2 and λ_{ext} 215.0 nm, $\Delta\varepsilon$ +113.6; A (= $\Delta\varepsilon_1 - \Delta\varepsilon_2$)= -251.8, due to exciton coupling between the ^1B_b transition of naphthalene and the intramolecular charge transfer or ^1L_a transition of two ethynylbenzene chromophores (Figure 4-4).

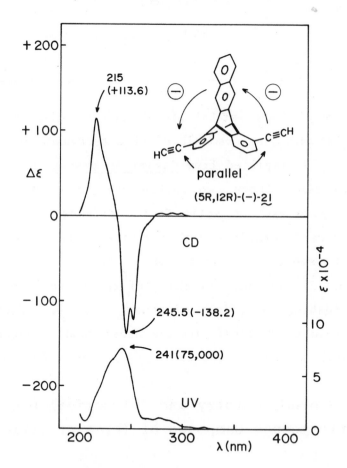

Figure 4-4. CD and UV spectra of (5$\underline{R}$,12$\underline{R}$)-(-)-1,15-diethynyl-5,12-dihydro-5, 12[1',2']benzenonaphthacene in ethanol. [Adapted from reference 24.]

Table 4-11.

322 (1,500)			
308 (1,400)			
— 278 (10,300)	290	(+3.5)	
268 (8,900)			
(— 241 (75,000)	245.5	(-138.2)	CD (24),(25)
	235	(0.0)	
	215	(+113.6)	Correl. (24)
O			
EtOH	EtOH		

21

In 21, the ^1B_b and intramolecular CT transition moments occupy ideal chiral positions for generating exciton coupling CD activity. Namely, combination of the long axes of the naphthalene and one ethynylbenzene moieties constitutes negative exciton chirality as shown in Figure 4-4. The same is true for the second combination of the naphthalene and the other ethynylbenzene. On the other hand, the third combination of two ethynylbenzene chromophores does not contribute to the CD activity because these two transitions are parallel to each other, the exciton chirality between them being nil. After all, compound 21 consists of two negative chiralities. Thus, the observed negative sign of the first Cotton effect leads to unequivocal and nonempirical determination of the (5R,12R) absolute configuration of (−)-21, which is in line with the chemical correlation results.[24]

As discussed in Chapter 11, the numerically calculated CD spectrum of (−)-21 based on the chiral exciton coupling mechanism is in excellent agreement with the observed curve, thus establishing the above assignment in a quantitative manner (see Figure 11-19).

The present chiral exciton coupling mechanism is confirmed by comparison of the CD spectra of the benzotriptycene diester (+)-22 and the hexahydrobenzotriptycene diester (-)-23 (Figure 4-5).   The diester (+)-22 exhibits typical and strong exciton Cotton effects of positive chirality, λ_{ext} 242.5 nm, $\Delta\varepsilon$ +151.1 and λ_{ext} 219.6 nm, $\Delta\varepsilon$ -177.8; A= +328.9.   On the other hand, the CD of

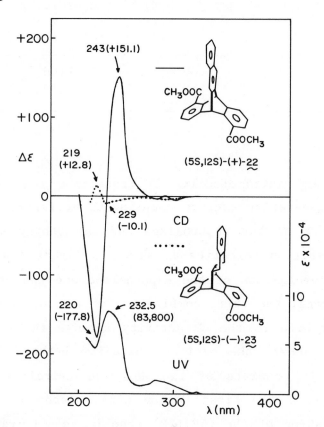

Figure 4-5. *Comparison of CD spectra of (5S,12S)-(+)-dimethyl 5,12-dihydro-5, 12[1',2']benzenonaphthacene-1,15-dicarboxylate 22 (solid line) and (5S,12S)- (-)-dimethyl 5,5a,6,11,11a,12-hexahydro-5,12[1',2']benzenonaphthacene-1,15-di- carboxylate 23 (dotted line).   The UV spectrum is that of 22.   [Adapted from reference 24.]*

Table 4-12.

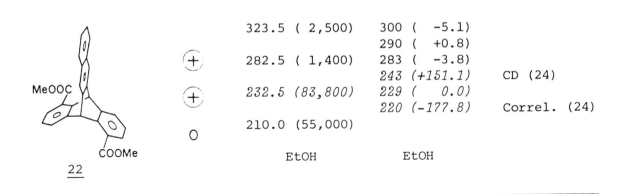

		323.5 (2,500)	300 (−5.1)	
			290 (+0.8)	
	(+)	282.5 (1,400)	283 (−3.8)	
			243 (+151.1)	CD (24)
	(+)	*232.5 (83,800)*	*229 (0.0)*	
			220 (−177.8)	Correl. (24)
	O	210.0 (55,000)		
		EtOH	EtOH	

22

(−)-23 is much weaker and noncharacteristic, λ_{ext} 229.0 nm, $\Delta\epsilon$ −10.1 and λ_{ext} 219.0 nm, $\Delta\epsilon$ +12.8; this is because (−)-23 lacks the naphthalene nucleus responsible for induction of exciton chirality.

Similarly, dialdehyde (+)-24 and dinitrile (+)-25 also exhibit typical exciton Cotton effects of positive chirality, which established the (5S,12S) absolute configuration (Figures 4-6 and 4-7).

Table 4-13.

			309.5 (−3.2)	
	(+)	282.0 (1,400)	291.5 (−4.2)	
			241.0 (+163.0)	CD (24)
	(+)	*235.5 (59,400)*	*234 (0.0)*	
			225.0 (−244.4)	Correl. (24)
	O	EtOH	EtOH	

24

For rigorous application of the CD exciton method, diester <u>22</u> and dialdehyde <u>24</u> are less suited than diethynyl compound <u>21</u> and dinitrile <u>25</u> because of
the conformational ambiguity around the single bond connecting the phenyl ring
to the substituents. However, the observed CD spectra are hardly affected by

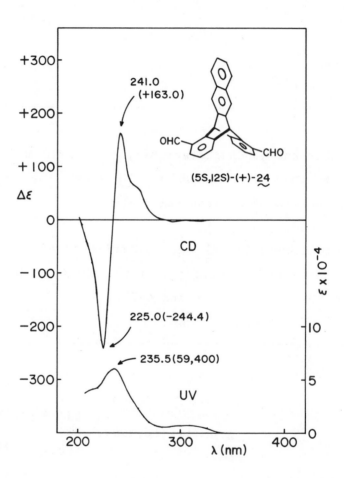

Figure 4-6. *CD and UV spectra of (5S-12S)-(+)-5,12-dihydro-5,12[1',2']benze-*
nonaphthacene-1,15-dicarbaldehyde in ethanol. [Adapted from reference 24.]

the conformational mobility of ester and aldehyde groups, as seen from Figures 4-5 and 4-6. The CD exciton chirality method is thus applicable to these systems in a straightforward manner.

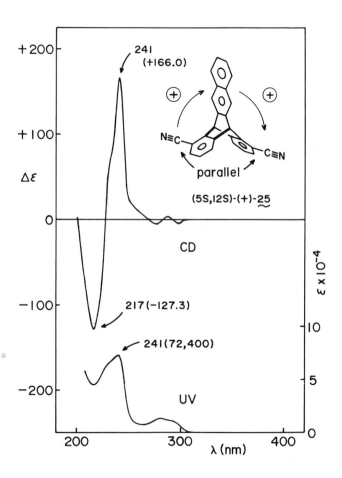

Figure 4-7. *CD and UV spectra of (5S,12S)-(+)-5,12-dihydro-5,12[1',2']benze-nonaphthacene-1,15-dicarbonitrile in ethanol. [Adapted from reference 24.]*

Table 4-14.

		323.5 (800)		
			298 (−4.5)	
	(+)		288 (+3.0)	
NC		282 (13,300)	277 (−5.0)	CD (24),(25)
	(+)	*241 (72,400)*	*241 (+166.0)*	
			228 (0.0)	Correl. (24)
	O		*217 (−127.3)*	
CN		EtOH	EtOH	
25				

In the case of chiral tribenzotriptycene,[24] (7$\underline{S}$,14$\underline{S}$)-(+)-7,14-dihydro-7,14
[1',2']naphthalenobenzo[$\underline{a}$]naphthacene ($\underline{26}$), the CD spectrum exhibits very
complex CD Cotton effects around 200-300 nm, which were entirely unexpected
from the simple exciton coupling mechanism (Figure 4-8). Since tribenzotriptycene (+)-$\underline{26}$ consists of two positive and one negative exciton chiralities
between three $^{1}B_{b}$ transitions of naphthalene moieties, two Cotton effects
of positive first and negative second signs were expected in the $^{1}B_{b}$
transition region, from the view point of chiral exciton interaction.
However, the observed CD spectrum showed four Cotton effects in the corresponding region.

The UV spectrum also reflects the complex electronic structural features
of compound $\underline{26}$. A new absorption band of medium intensity appears at 264.5 nm
(ε 35,800) which is not observable in the UV spectrum of naphthalene itself.
These phenomena imply that the interchromophoric homoconjugation effect is
contributing considerably to the CD and UV activities. In fact, as discussed
in Chapter 12, the SCF-CI-dipole velocity molecular orbital calculation
including the interchromophoric homoconjugation effect gave a good agreement

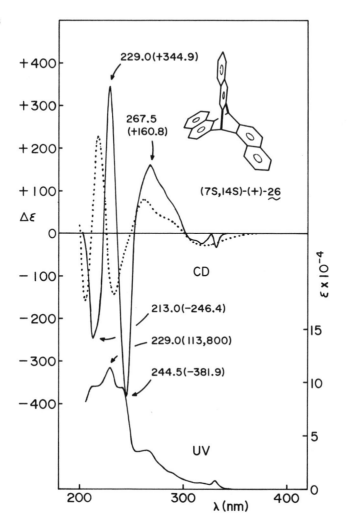

Figure 4-8. *CD and UV spectra of (7S,14S)-(+)-7,14-dihydro-7,14[1',2']naph-*
thalenobenzo[a]naphthacene in 0.2% dioxane/EtOH: UV, λ_{max} 331.0 nm (ε 7900),
264.5 (35,800), 241.5 (95,500), 229.0 (113,800), 214.0 (96,800); CD, λ_{ext}
331.5 nm ($\Delta\varepsilon$ -32.6), 317.0 (-24.6), 267.5 (+160.8), 244.5 (-381.9), 229.0
(+344.9), 213.0 (-246.4). Dotted line is the CD curve calculated by the SCF-
CI-DV molecular orbital method including the interchromophoric homoconjugation
between three naphthalenes: λ_{ext} 320.5 nm ($\Delta\varepsilon$ -29.0), 261.8 (+78.4), 232.5
(-142.0), 218.3 (+226.2), 205.7 (-157.3). [Adapted from reference 24.]

between the calculated and observed CD spectra (Figure 4-8). Thus, the absolute configuration of tribenzotriptycene (+)-26 can be assigned by the SCF-CI-DV molecular orbital calculation.

In order to determine the absolute stereochemistry of triple systems by applying the CD exciton chirality method in a reliable and nonempirical manner, it is necessary to choose ideal systems which satisfy the following requirements: (1) *combination of three chiralities of same sign or of two chiralities of same sign and one zero chirality*; namely, combination of plus/plus/plus or plus/plus/zero is more favorable than that of plus/plus/minus; (2) *negligible interchromophoric homoconjugation effect*. It should be emphasized that diethynyl (-)-21 and dinitrile (+)-25 benzotriptycenes composed of definite exciton chirality led to the first unambiguous and reliable chiroptical determination of absolute stereochemistry of chiral triptycenes.

4-5.   Exciton Coupling CD Spectra of Chiral Spirans

The following are examples of chiral spiro compounds with two aromatic chromophores that exhibit exciton split Cotton effects.

Table 4-15.

	284 (3,900)	286 (+1.9)	
	275 (3,400)	277 (+1.0)	
		269 (+6.4)	
		245 (+16.4)	
	237 (30,200)	*238 (0.0)*	CD (26)
		231 (-9.4)	
		211 (+4.8)	
	205 (81,300)	202 (-25.7)	
27	EtOH	EtOH	

Langer, et al.,[26] applied the chiral exciton coupling mechanism in order to interpret the CD spectra of chiral aromatic spiro compounds 27-30. All of these compounds exhibit relatively weak but typical exciton split Cotton effects of positive first and negative second signs in the region of intramolecular charge transfer or ^1L_a transition of mono-substituted benzene chromophores, as indicated in the tables. For example, dinitrile 27 shows split Cotton effects, λ_{ext} 245 nm, $\Delta\varepsilon$ +16.4 and 231 nm, $\Delta\varepsilon$ −9.4, in the $\pi\rightarrow\pi^*$ transition, λ_{max} 237 nm, ε 30,200. Since the benzonitrile moiety is unequivocal in conformation, the exciton chirality between the two long axes of benzonitrile moieties is unambiguously determinable; namely, the (R) absolute configuration depicted in 27 corresponds to positive exciton chirality. Thus, the absolute configuration of dinitrile was corroborated on the basis of the exciton chirality method.

Rigorously speaking, the exact direction of the $\pi\rightarrow\pi^*$ transition dipole moment of the acetyl, aldehyde and ester compounds is dependent on the conformation of these groups relative to the phenyl nucleus. However, in the present cases, both ortho positions are unsubstituted. It is therefore safe to conclude that the direction of transition dipoles all run along the long

Table 4-16.

	258 (30,900)	264 (+10.8)
		257 (0.0)
		247 (−11.8) CD (26)
		221 (+11.1)
		214 (−15.0)
	208 (45,700)	205 (+22.3)
28	EtOH	EtOH

	298 (+2.7)	
	267 (+11.6)	
262 (30,200)	260 (0.0)	
	252 (-12.9)	CD (26)
	222 (+8.1)	
	215 (-10.4)	
206 (39,800)	206 (+17.6)	
EtOH	EtOH	

287 (3,200)	288 (+1.4)	
278 (3,500)	279 (+1.2)	
	251 (+12.1)	
243 (26,900)	243 (0.0)	CD (26)
	235 (-10.5)	
	215 (+5.7)	
207 (57,500)	209 (-17.4)	
EtOH	EtOH	

axis; even if both ortho positions are unsymmetrically substituted, the direction of transition dipoles can be approximated by the long axis.   In any event, the (R) absolute stereochemistry with positive exciton chirality is unambiguously assignable to these compounds.

Shingu and coworkers[27] reported the conformational dependence of the exciton split Cotton effects in the system of 3,3'-di-tert-butyl-1,1'-spirobi [benz[e]indan]; three epimers 31, 32, and 33 exhibit typical split Cotton effects in the ^1B_b transition region, the amplitude of which reflects the

Table 4-17.

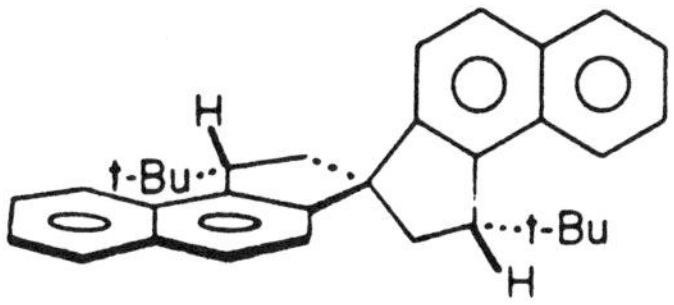

31

240 (+356.3)

224 (−409.1)   CD (27)

isooctane

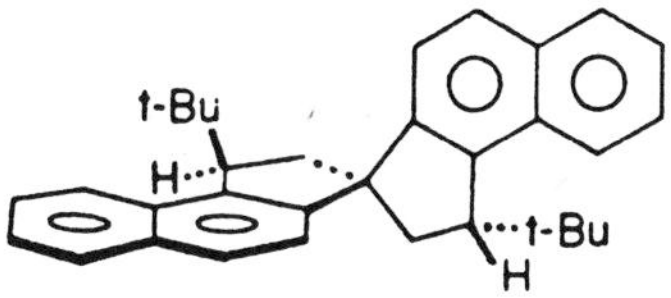

32

240.8 (+152.2)

223.5 (−191.2)   CD (27)

isooctane

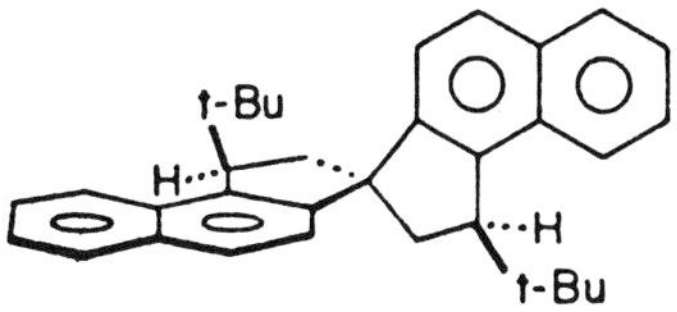

33

241   (+36.4)

221.3 (−50.1)   CD (27)

isooctane

---

conformational change of the dihedral angle between two naphthalene planes.

Table 4-18.

CD (28)

34

The absolute configuration of 3,3'-di-tert-butyl-1,1'-spirobi[benz[f]indan] (34) was also determined on the basis of the coupled oscillator mechanism.[28]

Table 4-19.

321.3 (2,700)	321.9 (+1.8)
313.4 (1,700)	314.2 (+0.8)
307.0 (2,200)	307.5 (+1.1)
283.2 (12,500)	285.0 (+4.4)
279.3 (12,800)	
274.1 (12,400)	
	230.2 (-961.5)
228.2 (172,700)	227.5 (0.0)
	221.6 (+567.1)
	207.0 (-27.3)

CD (16)

(1R,1'S,2S)-35

EtOH   EtOH

The CD spectrum of 2,2'-spirobi[benz[e]indan]-1,1'-diyl diacetate (35) shows very intense split Cotton effects of negative exciton chirality, from the sign of which the (1R,1'S,2S) absolute configuration was established.[16]

References

1. N. Harada, Y. Takuma, and H. Uda, J. Am. Chem. Soc. 98, 5408 (1976).

2. N. Harada, Y. Takuma, and H. Uda, Bull. Chem. Soc. Jpn. 51, 265 (1978).

3. N. Harada, Y. Takuma, and H. Uda, J. Am. Chem. Soc. 100, 4029 (1978).

4. N. Harada, Y. Takuma, and H. Uda, Bull. Chem. Soc. Jpn. 50, 2033 (1977).

5. S. Hagishita and K. Kuriyama, Tetrahedron 28, 1435 (1972).

6. R. Grinter and S. F. Mason, Trans. Faraday Soc. 60, 274 (1964).

7. S. F. Mason in "Optical Rotatory Dispersion and Circular Dichroism in Organic Chemistry," ed. G. Snatzke, (London: Heyden, 1967), Chapter 4.

8. S. F. Mason, R. H. Seal, and D. R. Roberts, Tetrahedron 30, 1671 (1974). (8a). N. Harada, unpublished data: the sample from Dr. S. Miyano, Tohoku University.

9. G. M. Badger, R. J. Drewer, and G. E. Lewis, J. Chem. Soc. 1962, p.4268.

10. H. Akimoto, T. Shioiri, Y. Iitaka, and S. Yamada, _Tetrahedron Lett._
     1968, p.97.

11. H. Akimoto and Y. Iitaka, _Acta Crystallogr. Series B_ 25, 1491
     (1969).

12. S. Yamada and H. Akimoto, _Tetrahedron Lett._ 1968, p. 3967.

13. H. Akimoto and S. Yamada, _Tetrahedron_ 27, 5999 (1971).

14. "IUPAC Tentative Rules for the Nomenclature of Organic Chemistry:
     Section E. Fundamental Stereochemistry" _J. Org. Chem._ 35, 2849
     (1970).

15. K. Harata and J. Tanaka, _Bull. Chem. Soc. Jpn._ 46, 2747 (1973).

16. N. Harada, unpublished data.

17. P. A. Browne, M. M. Harris, R. Z. Mazengo, and S. Singh, _J. Chem._
     Soc. C 1971, p. 3990.

18. H. E. Harris, M. M. Harris, R. Z. Mazengo, and S. Singh, _J. Chem._
     Soc., Perkin Trans. 2, 1974, p. 1059.

19. K. Mislow, M. A. W. Glass, R. E. O'Brien, P. Rutkin, D. H.
     Steinberg, J. Weiss and C. Djerassi, _J. Am. Chem. Soc._ 84, 1455
     (1962).

20. I. Hanazaki and H. Akimoto, J. Am. Chem. Soc. 94, 4102 (1972).

21. T. R. Govindachari, K. Nagarajan, P. C. Parthasarathy, T. G. Rajagopalan, H. K. Desai, G. Kartha, S.-M. L. Chen, and K. Nakanishi, J. Chem. Soc., Perkin Trans. 1 1974, p. 1413.

22. T. R. Govindachari, P. C. Parthasarathy, T. G. Rajagopalan, H. K. Desai, K. S. Ranachendran, and E. Lee, J. Chem. Soc., Perkin Trans. 1 1975, p. 2134.

23. M. Endo and H. Naoki, Tetrahedron 36, 2449 (1980).

24. N. Harada, Y. Tamai, Y. Takuma, and H. Uda, J. Am. Chem. Soc. 102, 501 (1980).

25. N. Harada, Y. Tamai, and H. Uda, J. Am. Chem. Soc. 102, 506 (1980).

26. E. Langer, H. Lehner, H. Neudeck, and K. Schlogl, Monatsh. Chem. 109, 987 (1978).

27. S. Imajo, A. Nakamura, K. Shingu, A. Kato, and M. Nakagawa, J. Chem. Soc., Chem. Commun. 1979, p. 868.

28. S. Imajo, A. Kato, K. Shingu, and H. Kuritani, Tetrahedron Lett. 22, 2179 (1981).

V. APPLICATION OF THE EXCITON CHIRALITY METHOD TO THE SYSTEM ALREADY POSSESSING ONE POLYACENE CHROMOPHORE

5-1. Chiral Exciton Coupling between Benzoate and Naphthalene Chromophores[1,2]

The present CD exciton chirality method can now be extended to interaction between different kinds of aromatic chromophores for which directions of the two interacting transitions are known. If a system in question already possesses such a chromophore, one can introduce an additional chromophore, e.g., benzoate chromophore, so that chiral exciton coupling between the two chromophores can be observed in the CD spectrum. From the sign of CD, the absolute stereochemistry of the original compound can be determined; it therefore

Table 5-1.

338	(2,800)	334 (+0.6)	
278	(6,500)	278 (+1.4)	
		235 (+34.0)	CD (1)
231.5	(77,600)	226 (0.0)	
		220 (−14.3)	X ray (3)
	EtOH	EtOH	

provides an extremely powerful method for determining absolute configurations or conformations of natural products, as exemplified in the following.

The extension was first demonstrated[1] with 17 -dihydroequilenin 3-methyl ether 17-benzoate (1, Figure 5-2). The free 17β-ol 4 exhibits Cotton effects centered at 330, 280, and 226 nm corresponding to the UV maxima of the methoxynaphthalene group. The ^1B_b transition (long axis) at 230 nm (Figure 5-1) is strongly coupled to the benzoate transition at 230 nm, and consequently the CD spectrum of 1 has two strong Cotton effects at 235 and 220 nm

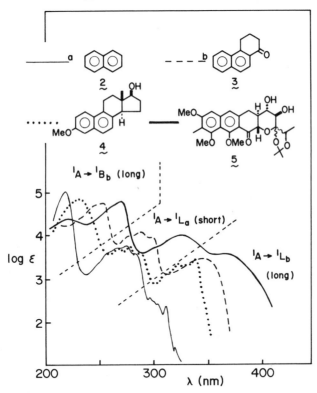

Figure 5-1. *UV spectra of pertinent naphthalenoid compounds in ethanol. (a) Adapted from reference 4. (b) Adapted from reference 5. [Adapted from reference 1.]*

(Figure 5-2). Significantly, the positive sign of the first Cotton effect is in agreement with the chirality between the long axes of the naphthalene moiety and the benzoate group (or to a first approximation, direction of the C(17)-O bond). The two other transitions, ^1L_a (short) at 280 nm and ^1L_b (long) at 330 nm, were not affected by the dipole-dipole coupling with the benzoate group because the bands are located far from the 230 nm benzoate transition; and their electric transition moments are relatively small, therefore the rotational strength (which is proportional to the electric transition moment) is also small.

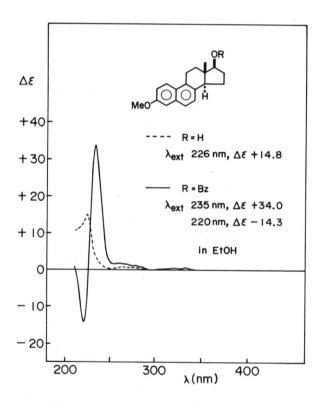

Figure 5-2. CD spectra of estra-1,3,5,7,9-pentaene-3,17β-diol 3-methyl ether and its benzoate in ethanol. [Adapted from reference 1.]

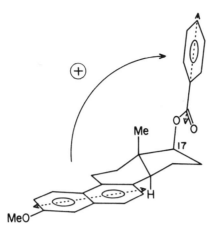

Figure 5-3. *Positive exciton chirality between long axes of benzoate and naphthalene moieties in the system of 3-methoxyestra-1,3,5,7,9-pentaen-17β-ol benzoate. [Adapted from reference 1.]*

The present method was next applied[1] to chromomycin A_3 belonging to the chromomycin and olivomycin group of antitumor antibiotics, the absolute configuration of which remained to be established.   The aglycone, chromomyci-

Table 5-2.

	362	(3,600)		ORD (1)
	330	(5,900)		
	316	(5,500)		CD (2)
	270.5	(42,500)	270 (-19.9)	
			242 (0.0)	Correl. (6)
	227.5	(26,000)	230 (+16.8)	
		EtOH	EtOH	

364 (4,300)		ORD (1)
331 (8,500)		
317 (7,700)		
	271 (-70.6)	CD (2)
264 (64,300)	*259 (0.0)*	
	250 (+34.0)	Correl. (6)
	220 (+9.9)	
EtOH	EtOH	

none (8), was isomerized to isochromomycinone (9) of established relative configuration, and this was finally converted into the benzoate 6 and p-methoxybenzoate 7.

The UV peaks due to the naphthalenoid moiety in glycol $\underline{5}$ can be readily assigned to various transitions by comparing the spectrum with related compounds (Figure 5-1).   The CD spectrum of glycol $\underline{5}$ shows weak negative and weak positive Cotton effects at 270 and 220 nm, respectively.   On the other hand, the CD of benzoate $\underline{6}$ exhibits strong typical exciton split Cotton effects of negative first and positive seconds signs; i.e., λ_{ext} 270 nm, $\Delta\varepsilon$

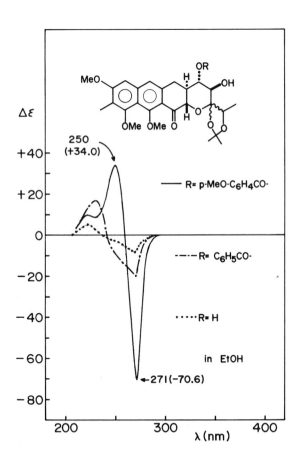

Figure 5-4.   CD   spectra of chromomycin derivatives in ethanol: (•••••), gly-col;   (–•–•–), benzoate; (————), p-methoxybenzoate.

−19.9 and 230 nm, Δε +16.8 (EtOH). This clearly indicates that the Cotton effects originate from the dipole-dipole interaction between the electric transition moments of the benzoate intramolecular charge transfer or ^1L_a transition (230 nm, ε 14,000) and the naphthalenoid ^1B_b transition (270 nm, ε 57,200). As the sign of the first Cotton effect is negative, the long axis transitions of the two chromophores interact as depicted in Figure 5-5 (negative chirality) and consequently, the absolute stereochemistry of chromomycinone was established as depicted by formula <u>8</u>.

The following results not only corroborate the above mentioned deductions but are also of practical value as they indicate that the interaction between two groups can be greatly enhanced by appropriately modifying the chromo-

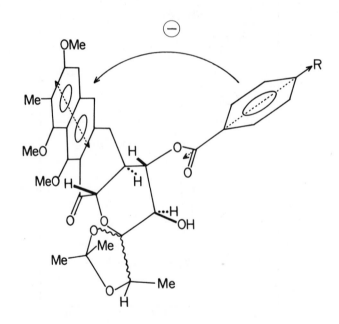

Figure 5-5. Negative exciton chirality in the system of chromomycin derivative benzoate. [Adapted from reference 1.]

phores.   The p-methoxybenzoate chromophore in 7 would be expected to interact much more efficiently with the naphthalenoid chromophore because its band
at 256 nm (ε 18,000) is located closer to the corresponding naphthalenoid ^1B_b
band, and according to the molecular exciton model the chromophoric interaction would be favored by similar excitation energies.   As shown in Figure 5-4,
this expectation was verified, and the apparent amplitude of the CD Cotton
effects is enhanced about three-fold: λ_{ext} 271 nm, $\Delta\varepsilon$ −70.6 and 250 nm, $\Delta\varepsilon$
+34.0(EtOH). The present conclusion was later confirmed by chemical corre-
lations.[6]

The absolute configuration of cervicarcin,[7] an antitumor antibiotic
produced by Streptomyces ogaensis, was also determined by application of the
exciton chirality method.   The UV spectrum of the alcoholic derivative 11

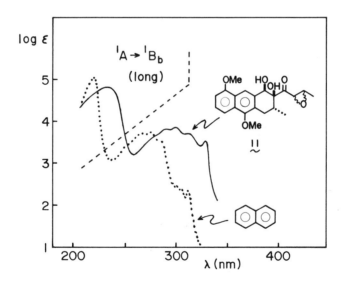

Figure 5-6. *UV spectra of cervicarcin derivative and naphthalene in ethanol.*

shows a pattern typical of α-substituted naphthalenes, in which the two longer wavelength transitions, ^1L_a (short axis) and ^1L_b (long axis), overlap with each other (Figure 5-6). Since the ^1B_b transition (long axis) at 234.5 nm has a very large absorption coefficient, strong Cotton effects due to interaction with the naphthalenoid chromophore were expected in the p-chlorobenzoate 10. In fact, the CD spectrum indicated this prediction to be correct; the CD spectrum of the p-chlorobenzoate showed two very strong exciton Cotton effects, λ_{ext} 242 nm, $\Delta\varepsilon$ +76.6 and 228 nm, $\Delta\varepsilon$ -46.2, while the alcohol exhibited only a simple Cotton effect, λ_{ext} 230 nm, $\Delta\varepsilon$ +6.0, as shown in Figure 5-7.

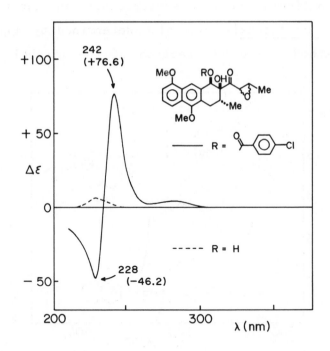

Figure 5-7. CD spectra of cervicarcin derivative 11 and its p-chlorobenzoate 10 in ethanol.

Table 5-3.

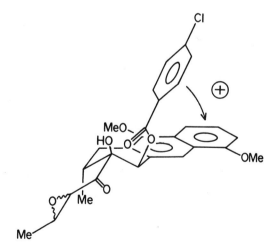

	301 (6,100)		
		242 (+76.6)	
	234.5 (62,000)	235 (0.0)	CD (7)
		228 (-46.2)	
	EtOH	EtOH	

The positive first Cotton effect sign indicates that the chirality between the long axes of the naphthalene and the p-chlorobenzoate chromophores
is positive, as depicted in Figure 5-8. Namely, the benzoyloxy group adopts a
β-configuration in which the chirality between two long axes is always positive, irrespective of the conformation of the cyclohexane ring.

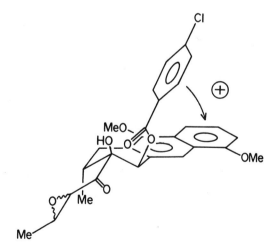

Figure 5-8. *Positive exciton chirality in the system of cervicarcin derivative p-chlorobenzoate. [Adapted from reference 7].*

Table 5-4.

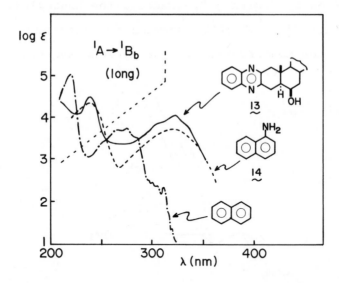

	322.5 (10,700)		
		243 (−23.6)	CD (2)
	239 (49,600)	*237 (0.0)*	
		231 (+24.6)	X ray (3)
	EtOH	EtOH	

A similar interaction is encountered in the quinoxaline derivative of 6β-p-chlorobenzoyloxy-5α-cholestane-2,3-dione (12);[2] the UV spectrum of the

Figure 5-9. *UV spectra of pertinent α-substituted naphthalenoid compounds in ethanol: quinoxaline derivative of 6β-hydroxy-5α-cholestane-2,3-dione, α-naphthylamine, and naphthalene.*

alcohol <u>13</u> is typical for an α-substituted naphthalene chromophore where the two longer wavelength bands, ^1L_a (short) and ^1L_b (long) transitions, mutually overlap (Figure 5-9). The shorter wavelength band at 239 nm, which is assignable to a ^1B_b (long axis) transition by comparison with other napththalenoid spectra (Figure 5-9), strongly interacts with the p-chlorobenzoate intramolecular charge transfer or ^1L_a band at 240 nm to give a negative first Cotton effect in accordance with the predicted negative chirality (Figure 5-10).

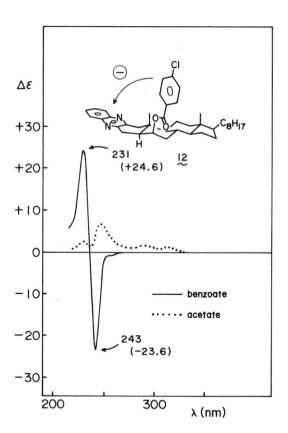

Figure 5-10. CD spectra of quinoxaline derivative of 6β-(p-chlorobenzoyloxy)-5α-cholestane-2,3-dione and its corresponding acetate in ethanol.

Cinchonine p-chlorobenzoate (15) and cinchonidine p-chlorobenzoate (16) represent interesting and unique examples because of the acyclic sec-hydroxyl groups[2].   The   two   alkaloids   have   opposing   configurations   at   two   chiral centers, and it was only after considerable confusion that the absolute con- figurations   were   finally   established.[8]   As   indicated   in   Figure   5-11,   the first Cotton effect signs are opposite; in each case the sign agrees with the chirality between the long axes of the quinoline and p-chlorobenzoate chromo- phores adopting the preferred conformation.

Table 5-5.

		318 (7,900)		
	$\ominus$	238 (51,900)	241 (−19.1) 234 (0.0) 231 (+2.4)	CD (2) X ray (8)
		0.1 N HCl	0.1 N HCl	
15				
	$\oplus$		244 (+15.9) 234 (0.0) 231 (−2.2)	CD (2)
			0.1 N HCl	
16				

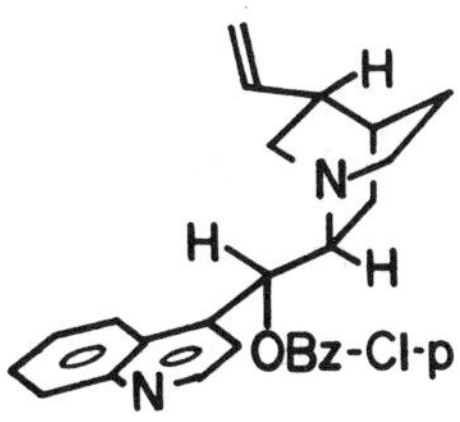

15 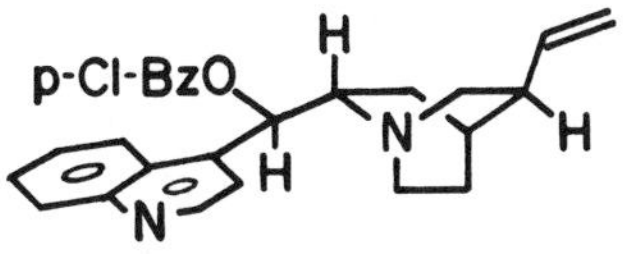 **16**

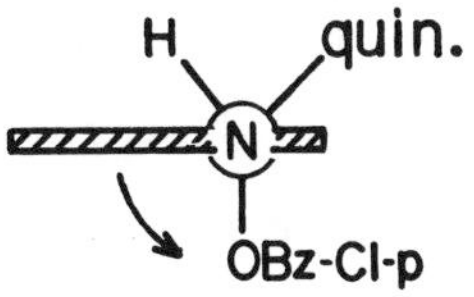

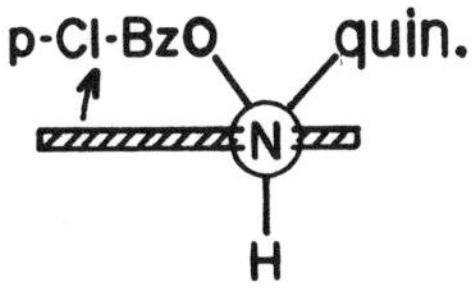

λ_{ext} 241 nm, $\Delta\varepsilon$ −19.1 λ_{ext} 244 nm, $\Delta\varepsilon$ +15.9

231 nm,   + 2.4   231 nm,   − 2.2

Figure 5-11. Exciton chiralities in the systems of cinchonine and cinchonidine p-chlorobenzoates. [Adapted from reference 2].

Table 5-6.

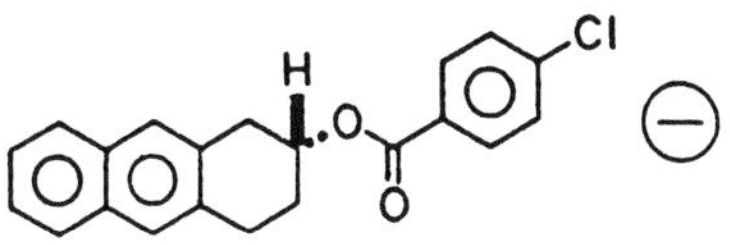

17

235 (−32.3)

225 (+32.8)   CD (9)

10% dioxane/
MeOH

Koreeda and coworkers[9] applied the exciton chirality method to 1,2,3,4-tetrahydro-2-anthryl p-chlorobenzoate (17), in order to determine the absolute configuration of 1,2-dihydroanthracene-trans- and -cis-1,2-diols produced from anthracene by mammals and bacteria.   1,2,3,4-Tetrahydroanthracene and p-chlorobenzoate chromophores exhibit strong $\pi \rightarrow \pi^*$ transitions at 230 nm (^1B_b, ε 100,000) and 240 nm (intramolecular charge transfer or ^1L_a, ε 24,000), respectively.   The observed CD spectrum of the p-chlorobenzoate 17 showed a typical exciton interaction pattern of two Cotton effects with opposite signs, λ_{ext} 235 nm, $\Delta\varepsilon$ −32.3 and 225 nm, $\Delta\varepsilon$ +32.8.   The negative sign of the first Cotton effect led to the (2R) absolute configuration of negative exciton chirality.   The absolute stereochemistry of other metabolites was determined by chemical correlation with the present (+)-1,2,3,4-tetrahydro-2-anthrol.

Endo, et al.,[10] determined the absolute configuration of germichrysone and torosachrysone, tetrahydroanthracenes isolated from Cassia torosa and Cassia singueana by use of the coupled oscillator method.   The CD spectrum of germichrysone dimethyl ether benzoate (18) exhibits strong positive first and negative second exciton Cotton effects, while no significant Cotton effects were observable in the CD spectrum of germichrysone itself.   The ^1H NMR coupling constant data ($J_{2e,3e}$= 5.8 Hz; $J_{2a,3e}$= 4.5 Hz) showed the benzoate

Table 5-7.

---

	373 (7,700)		CD (10)
	320 (3,900)		
	310 (5,200)		
	262 (63,500)	258 (+102.1)	X ray (11)
	224 (42,700)	229 (−138.6)	
18	MeOH	MeOH	

---

366 (7,600)		
332 (5,200)		
320 (5,000)		
266 (60,300)	*262 (-62.2)*	CD (10)
225 (36,900)	*229 (+87.7)*	
MeOH	MeOH	

group to be axial.   The observed split Cotton effects of positive exciton chirality then leads to the (3R) absolute configuration of germichrysone.   The present results are in line with recent X-ray crystallographic data.[11]   In the case of torosachrysone, it was possible to benzoylate the tertiary alcohol group.   The CD spectrum of the dimethyl ether benzoate 19 shows exciton split Cotton effects of negative chirality, which leads to (3S) absolute configuration of torosachrysone.   In these cases, the p-methoxybenzoate chromophore would be more favorable than the unsubstituted benzoate because of the proximity between UV λ maxima of the two chromophores.

References

1. N. Harada, K. Nakanishi, and S. Tatsuoka, J. Am. Chem. Soc. 91, 5896 (1969).

2. N. Harada and K. Nakanishi, Acc. Chem. Res. 5, 257 (1972).

3. Absolute configurations of steroidal compounds are established by the X-ray Bijvoet method: see W. Klyne and J. Buckingham, Atlas of Stereochemistry (London: Chapman and Hall 1978) vol. 1, p. 121-26; vol. 2, pp. 63-64.

4. W. V. Mayneord and E. M. F. Roe, Proc. Roy. Soc (London) A152, 299 (1935).

5. R. Hursgen and U. Rietz, Tetrahedron 2, 271 (1958).

6. Chemical correlation of related compound: Yu. A. Berlin, M. N. Kolosov, and L. A. Piotrovich, Tetrahedron Lett. 1970, p. 1329.

7. S. Marumo, N. Harada, K. Nakansihi, and T. Nishida, J. Chem. Soc., Chem. Commun. 1970, p. 1693.

8. X ray of quinidine: O. L. Carter, A. T. McPhail, and G. A. Sim, J. Chem. Soc. A 1967, p. 365.

9. M. N. Akhtar, D. R. Boyd, N. J. Thompson, M. Koreeda, D. T. Gibson, V. Mahadevan, and D. M. Jerina, J. Chem. Soc., Perkin Trans. 1, 1975, p. 2506.

10. M. Endo and H. Naoki, <u>Tetrahedron</u> <u>36</u>, 2449 (1980).

11. H. Noguchi, U. Sankawa, and Y. Iitaka, <u>Acta Crystallogr., sect. B</u> <u>34</u>, 3273 (1978).

VI. APPLICATION OF THE EXCITON CHIRALITY METHOD TO CONJUGATED ENONES, ESTERS, ETC.

6-1. Application to Bis(enone) System[1]

The exciton chirality method can be extended to the $\pi \rightarrow \pi^*$ transition of conjugated polyene, enone, ester, lactone, and amide systems, as in the case of dibenzoate and bis(polyacene) chromophores. The $\pi \rightarrow \pi^*$ transitions of these systems are polarized along the long axes of the chromophores, and their UV absorptions are sufficiently intense ($\varepsilon \simeq 15,000$) for the observation of chiral exciton coupling.

Table 6-1.

$\underline{1}$ $\oplus$	255 (11,700)	330 (−2.5) 266 (+10.4) 242 (−9.5)	CD (1)
	EtOH	EtOH	

Quassin, a decanortriterpene and plant bitter principle, has two enone moieties with the π→π* transition at λ_{max} 255 nm, ε 11,700 (EtOH). The CD spectrum shows typical exciton split Cotton effects of positive first and negative second signs, λ_{ext} 266 nm, Δε +10.4 and 242 nm, Δε -9.5 (EtOH), in the π→π* transition region (Figure 6-1).[1] From the positive sign of the observed first Cotton effect, the absolute configuration of positive exciton chirality was assigned to quassin as depicted in 1.

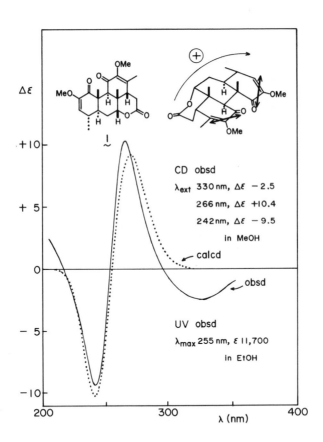

Figure 6-1. CD spectra of quassin: (————) observed in methanol, (••••••) calculated. [Adapted from reference 1.]

Moreover, the excellent agreement seen between the observed and calculated CD curves of quassin (Figure 6-1) not only establishes the absolute configuration of quassin itself but also proves that the observed CD of quassin around 250 nm is due to exciton coupling.

This conclusion was consistent with the previous absolute configurational assignment of quassin which was based on biosynthetic considerations and X-ray analysis of biogenetically related compounds.[2]

Although quassin shows a negative $n \to \pi^*$ Cotton effect at 330 nm, the optical rotation at the sodium D-line (589 nm) is positive: $[\alpha]_D$ +34.5° (c 5.09, $CHCl_3$).[3]   Therefore, the D-line rotation, which is simply the reading at the D-line taken from ORD curves, is governed by the intense positive Cotton effect at 266 nm. Similar phenomenon is also observed in the case of abscisic acid (see section 6-2).

6-2. Absolute Configuration of Abscisic Acid

The absolute configuration of (+)-cis-abscisic acid remained obscure in spite of its important role as a plant-growth regulator. Although Cornforth, et al.,[4] had assigned an (R) configuration to (+)-cis-abscisic acid by application of the Mills' rule,[5] Burden, et al.,[6] concluded that the absolute stereochemistry of either abscisic acid or violaxanthin was incorrect from chemical correlation studies with violaxanthin. Shortly afterwards, Isoe, et al.,[7] and Oritani, et al.,[8] independently indicated, on the basis of chemical correlations, that the absolute configuration of abscisic acid should be revised and expressed by the (S) configuration. Ryback[9] also arrived at the same conclusion by direct chemical correlation of (+)-cis-abscisic acid with (S)-malic acid.

Table 6-2.

(+)	234 (21,300) MeOH	320 (-2.3) 242 (+38.4) 208 (-30.2) MeOH	CD (10) Correl. (7), (8),(9),(12)

$\underline{2}$

The absolute configuration of abscisic acid was also established independently from CD data. The optically active dehydrovomifoliol ($\underline{2}$) having two enone groups in the ring and side chain was prepared.[10] The $\pi \rightarrow \pi *$ transition of the two enone groups, λ_{max} 234 nm, ε 21 300, couple with each other to generate exciton split Cotton effects of positive first and negative second signs, λ_{ext} 242 nm, $\Delta\varepsilon$ +38.4 and 208 nm, $\Delta\varepsilon$ -30.2. Therefore, the (S) absolute configuration with positive exciton chirality was assignable to compound $\underline{2}$. Since compound $\underline{2}$ was convertible to (+)-cis-abscisic acid, the absolute configuration of natural abscisic acid was determined to be S.

Similar exciton coupling Cotton effects were observable in the case of (+)-cis- and -trans-abscisic acids having interacting enone and diene-carboxylic acid groups.[10,13] Both compounds exhibit exciton Cotton effects of positive first and negative second signs as shown in Table 6-3. These results also

Table 6-3.

			317 (-2.3)	
			261 (+34.5)	CD (10),(13)
		245 (24,800)		
			229 (-28.0)	Correl. (7)
				(8),(9),(12)
(S)-(+)-natural		MeOH	MeOH	
3				

			323 (-2.2)	
			254 (+25.5)	CD (10)
		241 (28,400)		
			221 (-12.6)	Correl. (7)
				(8),(9),(12)
4		MeOH	MeOH	

led to the same conclusion as described above. Furthermore, the (S) absolute configuration of natural abscisic acid was established again by applying the exciton chirality method to the enone-benzoate system, as discussed in Chapter 7.

A calculation of the ORD spectrum of (+)-trans-abscisic acid based on the molecular exciton theory also made it possible to assign the (S) configuration to the acid.[11]

Later, Mori[12] also confirmed the reversal of the absolute configuration of abscisic acid described above, by chemically correlating the acid with grasshopper ketone of known absolute configuration.

Similar to the case of quassin (section 6-1), the sign of $[\alpha]_D$ of (+)-cis- and trans-abscisic acids is opposite to that of the n→π* Cotton effect at the longest wavelength; it however agrees with that of the first extremum of split Cotton effects at the π→π* transition (see Table 6-3).

6-3.   Exciton Cotton Effects of Bis(Conjugated Polyene) Systems

Exciton coupling Cotton effects are also observable in the case of interaction between two diene-carboxylic acid chromophores; i.e., bis(sorbate) system.[14]

Table 6-4.

268 (-20.4)	
256 (0.0)	CD (14)
245 (+16.5)	
5% dioxane/	
EtOH	

5

270 (+20.2)	
255 (0.0)	CD (14)
240 (-14.4)	
5% dioxane/	
EtOH	

6

The CD spectra of steroidal glycol bis(sorbates) exhibit typical exciton split Cotton effects in the region of the π→π* transition, as shown in Table 6-4 and Figure 6-2.   The amplitude of the Cotton effects is comparable to that of regular dibenzoates. However, the sorbate chromophore is less favorable than the benzoate chromophores due to less symmetrical structure and conformational ambiguity around the alcoholic C–O bond.

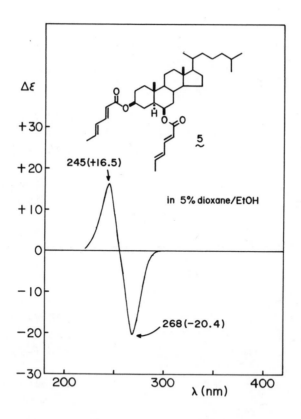

Figure 6-2. *CD spectrum of 5α-cholestane-3β,6β-diol bis(sorbate).*

Table 6-5.

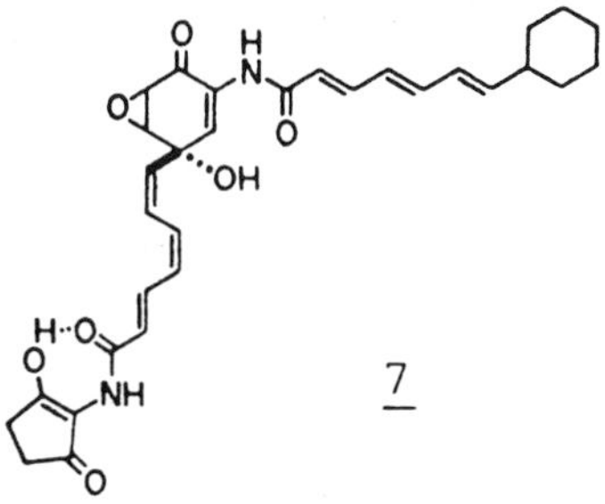

339 (+39.0)

301 (-49.2)

CD (15)

MeOH

7

$\oplus$

This is an interesting example of chiroptical determination of absolute configurations of a conjugated polyene-amide system.   Kakinuma, et al.,[15] determined the absolute configuration of asukamycin, an antibiotic produced by Streptomyces nodosus subs. asukaensis, from the CD spectral data.   Asukamycin (7) exhibits intense split Cotton effects due to chiral exciton coupling between the two polyene-amide chromophores in the $\pi \rightarrow \pi *$ transition region: λ_{ext} 339 nm, $\Delta\varepsilon$ +39.0 and 301 nm, $\Delta\varepsilon$ -49.2.   From the positive sign of the first Cotton effect, a positive exciton chirality between the two polyene-amide groups was deduced.   The (S) configuration at C-4 position was thus determined.

References

1. M. Koreeda, N. Harada, and K. Nakanishi, J. Am. Chem. Soc. 96, 266 (1974).

2. X-ray of simarolide: W. A. C. Brown and G. A. Sim, Proc. Chem. Soc., London 1964, p. 293.

3. E. London, A. Robertson, and H. Worthington, J. Chem. Soc. 1954, p. 3431.

4. J. W. Cornforth, W. Draber, B. V. Milborrow, and G. Ryback Chem. Commun. 1967, p. 114.

5. J. A. Mills, J. Chem. Soc. 1952, p. 4976.

6. R. S. Burden and H. F. Taylor, Tetrahedron Lett. 1970, p. 4071.

7. S. Isoe, S. B. Hyeon, S. Katsumura, and T. Sakan, Tetrahedron Lett. 1972, p. 2517.

8. T. Oritani and K. Yamashita, Tetrahedron Lett. 1972, p. 2512.

9. G. Ryback, Chem. Commun. 1972, p. 1190.

10. M. Koreeda, G. Weiss, and K. Nakanishi, J. Am. Chem. Soc. 95, 239 (1973).

11. N. Harada, J. Am. Chem. Soc. 95, 240 (1973).

12. K. Mori, Tetrahedron Lett. 1973, p. 2635.

13. T. Oritani, K. Yamashita, and H. Meguro, Agr. Biol. Chem. (Japan) 36, 885 (1972).

14. N. Harada, unpublished data.

15. K. Kakinuma, N. Ikekawa, A. Nakagawa, and S. Omura, J. Am. Chem. Soc. 101, 3402 (1979).

VII. APPLICATION OF THE EXCITON CHIRALITY METHOD TO THE SYSTEM POSSESSING ONE ENONE OR DIENE CHROMOPHORE

7-1. Chiral Exciton Coupling between Benzoate and Enone Chromophores[1,2]

The $\pi \to \pi^*$ transition of the enone system around 230–260 nm (ε 7000–15,000) lies close to the benzoate intramolecular charge transfer or ^1L_a transition at 230 nm ($\varepsilon \simeq 14,000$). Accordingly, the absolute configurations of molecules containing these two chromophores in close proximity can be determined from the signs of the split CD curves.

Table 7-1.

246 (+16.7)

221 (−21.6)

CD (1)

MeOH

1

The first example is the case of 7-oxocholest-5-en-3β-ol p-chlorobenzoate
(1), which shows positive first and negative second Cotton effects, λ_{ext}
246 nm, Δε +16.7 and 221 nm, Δε -21.6, due to the coupling between enone and
p-chlorobenzoate moieties.[1] Since the corresponding acetate 2 simply exhibits
a negative Cotton effect around 210 nm (Figure 7-1), it is clear that exciton
coupling governs the CD Cotton effects of the p-chlorobenzoate around 240 nm.
The positive sign of the observed first Cotton effect is in accord with the
positive exciton chirality between the long axes of benzoate and enone
moieties (Figure 7-2).

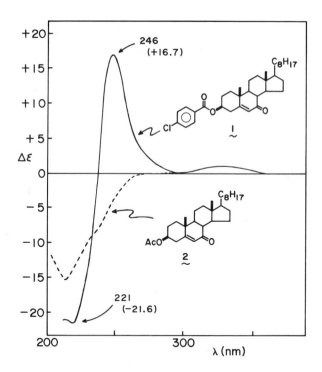

Figure 7-1. *CD spectra of 7-oxocholest-5-en-3β-ol   p-chlorobenzoate   (solid
line) and 3-acetate (dotted line) in methanol. [Adapted from reference 1].*

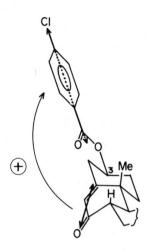

Figure 7-2. *Positive exciton chirality in the system of 7-oxocholest-5-en-3β-ol p-chlorobenzoate.* *[Adapted from reference 1.]*

Table 7-2.

	237 (+7.3)	CD (1)
	217 (-18.9)	
3	MeOH	

It is advisable to choose a para-substituted benzoate having its long axis transition maximum close to the enone π→π* maximum because of the greater magnitude of the resulting split CD Cotton effects. For example, the unsubstituted benzoate **3** shows weaker exciton Cotton effects than those of p-chlorobenzoate $\underline{1}$. This phenomenon is explicable as follows: λ_{max} of the $\pi \rightarrow \pi^*$ transition of $\underline{p}$-chlorobenzoate chromophore (240 nm, ε 21,400) is almost identical with that of enone moiety (239 nm, ε 12,000), and this leads to enhancement of the harmonious exciton coupling between enone and $\underline{p}$-chlorobenzoate transitions.   On the other hand, in $\underline{3}$, there is a gap of 10 nm between λ_{max} of enone and unsubstituted benzoate (229.5 nm, ε 15,300);   the   exciton coupling is less favorable and the split CD amplitude is smaller (see section 3-1-D).

Table 7-3.

241 (−15.5)		
		CD (1)
211 (+21.1)		
	MeOH	

247 (−24.4)		
		CD (1)
224 (+23.0)		
	MeOH	

259 (−16.8)

238 (+21.4)   CD (1)

MeOH

6

OMe

The above mechanism was verified by compounds 4, 5, and 6. In the case of exciton coupling with the π→π* transition (241 nm, ε 16,600) of the enone group in ring A, p-chlorobenzoate is superior to unsubstituted and p-methoxy (257 nm, ε 20,400) benzoates for the same reason discussed above. In fact, p-chlorobenzoate 5 exhibits more intense exciton split Cotton effects than 4 and 6 (see Table 7-3).

Table 7-4.

250 (+13.1)

230 (−3.3)   CD (1)

MeOH

7

Similarly, compound 7 also exhibits exciton split Cotton effects.

Table 7-5.

315 (−1.2)
243 (−10)

223 (+16)   CD (2)

EtOH

8

318 (−1.5)
240 (+22)   CD (2)

EtOH

9

Delaroff, et al.,[2] reported exciton Cotton effects due to coupling between the ring A enone and ring D benzoate. Note that the coupled oscillator mechanism is still effectively operating in such a remote distance.

7-2.   Application for Determining Absolute Stereochemistry of Abscisic Acid[3]

Chiral exciton coupling of enone-benzoate system was also utilized for elucidating the absolute configuration of (+)-cis-abscisic acid; hydrogenation of
peroxide 10 and optical resolution of the resulting racemic cis-diol in the
form of MTPA ester by high speed liquid chromatography gave enantiomeric
cis-diols 11 and 13.   Diol 11 was benzoylated and oxidized to afford the
enone-benzoate derivative 12, while the other diol 13 was converted into
natural (+)-cis-abscisic acid.

The enone-benzoate derivative 12 exhibits typical exciton split Cotton
effects of negative first and positive second signs (Table 7-6).   Hence, the
absolute configuration with negative exciton chirality between enone and
benzoate π→π* transitions was assigned to 12.

Table 7-6.

	231 (19,800)	336 (+0.7)	CD (3)
12		238 (-19.0)	
		219 (+3.2)	
	MeOH	MeOH	

7-3. Chiral Exciton Coupling between Benzoate and Conjugated Diene Chromophores

The absolute configuration of periplanone B, a sex pheromone of the American cockroach, Periplaneta americana, was determined by applying the exciton chi-

Table 7-7.

	229 (-)	CD (4)
15	227 (0)	
	223 (+)	
	hexane	

rality method to a derivative with conjugated diene and benzoate chromophores, as follows.[4]   The relative configuration was determined by a total synthesis[5] and finally by the X-ray crystallographic study[4] of synthetic and racemic periplanol B (10-equatorial-OH as shown in Figure 7-3a).   In order to determine the absolute configuration, the MTPA (α-methoxy-α-trifluoromethyl-phenylacetic acid) ester of synthetic periplanol B was resolved by LC separation, the MTPA esters were hydrolysed, and the two resolved alcohols were oxidized to the ketones (periplanone B and its enantiomer) for bioassay.   Both alcohols were converted into their benzoates in the expectation that the benzoate 229 nm absorption would couple with the diene 227 nm band. However, this was not the case (Figure 7-3a shows the CD of one enantiomer); the 225nm diene

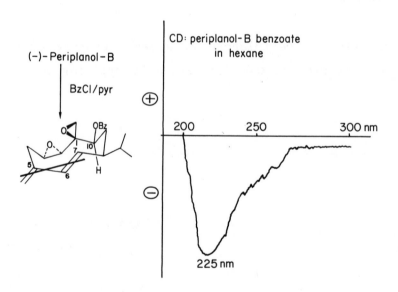

Figure 7-3a. CD spectrum of periplanol B benzoate in hexane.   [Adapted from reference 4].

Cotton effect simply overrides an uncoupled benzoate Cotton effect. The resolved biologically active periplanone B (50 µg) was therefore reduced with sodium borohydride to give 10-epi-periplanol B with an axial 10-hydroxyl group, which was benzoylated. As illustrated in Figure 7-3b, the benzoate now exhibits negative first and positive second split Cotton effects arising from the exciton coupling between diene and benzoate $\pi \rightarrow \pi^*$ transitions. The absolute configuration depicted in the figure was assigned to the benzoate from the negative sign of the first Cotton effect. The result was in agreement with that derived from the diene helicity rule[6] and the allylic chirality method.[7]

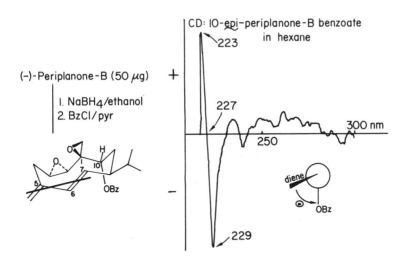

Figure 7-3b. CD spectrum of 10-epi-periplanol B benzoate in hexane. [Adapted from reference 4.]

Table 7-8.

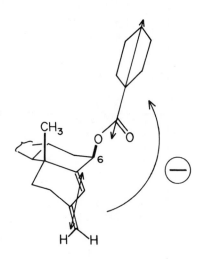	235.0 (33,800)	242.0 (−30.8) 233.7 (0.0) 225.0 (+39.1)	CD (8)
16	EtOH	EtOH	

The chiral exciton coupling between conjugated diene and benzoate chromophores is clearly observable in 3-methylenecholest-4-en-6β-ol benzoate as illustrated in Figure 7-5.[8]   In this case, the π→π*   transition   of   transoid

Figure 7-4. *Negative exciton chirality between conjugated diene and benzoate chromophores in the system of 3-methylenecholest-4-en-6β-ol benzoate.*

conjugated diene at 239 nm couples with the benzoate's π→π* transition at 230 nm.   As shown in Figure 7-4, since the two transitions constitute a negative exciton chirality, split Cotton effects of negative chirality are predicted. This is verified in Figure 7-5.

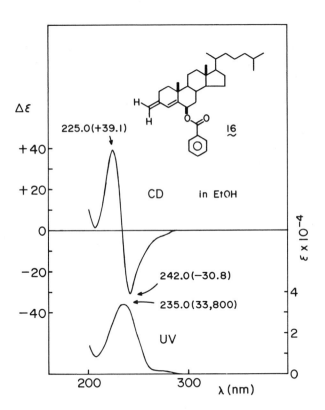

Figure 7-5.   *CD and UV spectra of 3-methylenecholest-4-en-6β-ol   benzoate   in ethanol.*

7-4. A Method for Determining Absolute Configuration of Allylic Alcohols on the Basis of Chiral Exciton Coupling between Double Bond and Benzoate Chromophores [9]

As will be discussed in Chapter 10, the exciton chirality in nondegenerate systems composed of two different chromophores, as in the case of degenerate systems with two identical chromophores, is also theoretically defined by the quadruple product, $\vec{R}_{ij} \cdot (\vec{\mu}_{i0a} \times \vec{\mu}_{j0b}) V_{ij}$:   thus, if the quadruple product is positive in a system, the first Cotton effect at longer wavelengths is positive while the second Cotton effect at shorter wavelengths is negative. The feature of the chiral exciton coupling in nondegenerate systems was theoretically and experimentally exemplified by the CD spectra of steroidal benzoate/<u>para</u>-substituted benzoate systems (Figures 3-14 and 3-15).   It was emphasized in these cases that the exciton coupling mechanism still governs the CD power of the 230 nm-310 nm system, the Cotton effects of which are completely separated in wavelength from each other.

The concept of the chiral exciton coupling in nondegenerate systems is extendable to allylic benzoates; i.e., double bond-benzoate systems. The benzoate chromophore exhibits an allowed $\pi \rightarrow \pi^*$ intramolecular charge transfer band at 230 nm, and the C–C double bond chromophore also shows an allowed $\pi \rightarrow \pi^*$ transition around 195 nm. [10] Both $\pi \rightarrow \pi^*$ transitions are polarized along the long axes of the chromophores, respectively. Therefore, if the two long axes of benzoate and double bond chromophores consist of positive exciton chirality, i.e., positive quadruple product or right-handed screwness, the first Cotton effect at longer wavelengths (230 nm benzoate Cotton effect) is positive, while the second Cotton effect at shorter wavelengths (195 nm double bond $\pi \rightarrow \pi*$ Cotton effect) is negative (Figure 7-6). On the other hand, if allylic benzoate consists of negative exciton chirality, the 230 nm benzoate Cotton effect is negative.

In addition to the intense $\pi \rightarrow \pi*$ (or $\pi_x \rightarrow \pi_x*$) transition, a C–C double bond chromophore has additional weak $\pi \rightarrow \sigma*$ and $\pi_x \rightarrow \pi_y*$ transitions.[11] Moreover since the olefinic chromophore is also asymmetrically perturbed by allylic substituents[12] and/or by skeletal strain of the double bond, overlap of

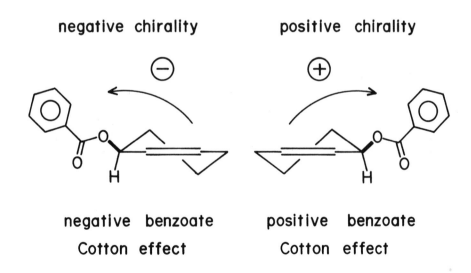

negative chirality   positive chirality

negative benzoate   positive benzoate
Cotton effect   Cotton effect

Figure 7-6. Exciton chirality of allylic benzoates and sign of benzoate Cotton effects.

additional Cotton effects complicates the CD curve of the double bond chromophore. On the other hand, since the benzoate chromophore is strain-free, the
230 nm benzoate transition is mainly affected by the 195 nm allowed $\pi \to \pi^*$ transition of the double bond as a result of the exciton coupling mechanism in
nondegenerate systems (Figure 7-7).   The present method is thus based on
theoretically sound grounds.

The fact that the exciton chirality between the benzoate and double bond
chromophores is independent of conformational change is significant from the
view point of actual application of the present method.   For instance, in the
case of (R)-2-cyclohexen-1-ol benzoate, the cyclohexene ring can adopt two

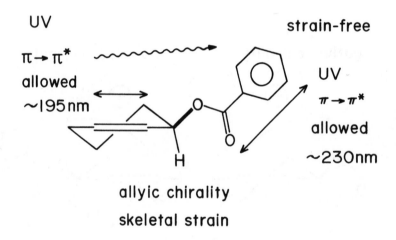

Figure 7-7. The chiral exciton coupling mechanism in allylic benzoate, a
nondegenerate system.

half-chair conformations as depicted in Figure 7-8.   In both conformers, the
exciton chirality between the benzoate and double bond chromophores is positive irrespective of conformational change of the cyclohexene ring.   Thus, a
positive benzoate Cotton effect leads to (<u>R</u>) configuration in a straightforward manner.   This fact contrasts favorably with the case of the cyclohexenone
system, in which the enone helicity is reversed from one half-chair conformer
to the other (Figure 7-8).

Exciton Chirality:
independent of conformational change

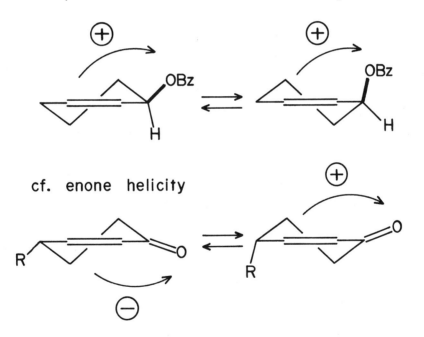

Figure 7-8. The sign of the exciton chirality of allylic benzoate systems is
independent of the conformational change of cyclohexene ring, while the sign
of enone helicity depends on the conformation and is reversed.

The first example is the case of cholest-4-en-3β-ol benzoate (<u>17</u>) in Figure 7-9. In the region of the π→π* intramolecular charge transfer transition at 228.9 nm (ε 16,400), the CD spectrum exhibits a negative Cotton effect, λ_{ext} 229.5 nm, Δε -8.72. The negative sign of the CD Cotton effect is in accordance with the negative exciton chirality between the benzoate and double bond chromophores (Figure 7-9).

The negative CD ellipticity observed below 210 nm does not follow the theoretical expectation that the signs of first and second Cotton effects

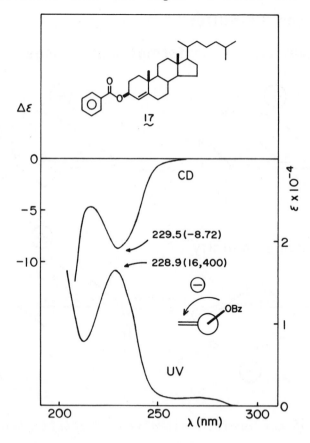

Figure 7-9.   CD and UV spectra of cholest-4-en-3β-ol benzoate in ethanol.

should be opposite to each other. This intense ellipticity is probably due to the participation of the benzenoid [1]B transition (around 200 nm) of the benzoate chromophore, and/or the CD activity of the double bond moiety discussed above.

As in the case of the dibenzoate chirality method, para-substituted benzoates can also be used in the present method. Figure 7-10 shows the effect of para-substituents on the CD and UV spectra of cholest-4-en-3β-ol   benzoate

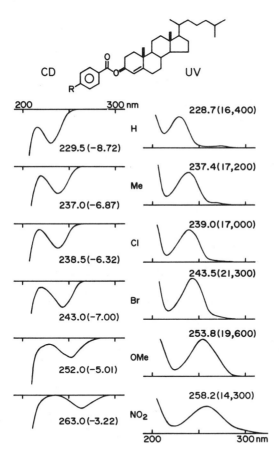

Figure 7-10. *CD and UV spectra of cholest-4-en-3β-ol   para-substituted benzoates in ethanol.*

system.   Both strongly electron–donating and –withdrawing groups cause a red–
shift of the π→π* intramolecular charge transfer band.   As the wavelength
separation between the benzoate and double bond π→π* transitions increases,
the Δε value of the benzoate Cotton effects decreases as expected from theory.
However, since the sign of benzoate Cotton effects remains unchanged, all of
the para–substituted benzoate chromophores shown in Figure 7–10 can be used.

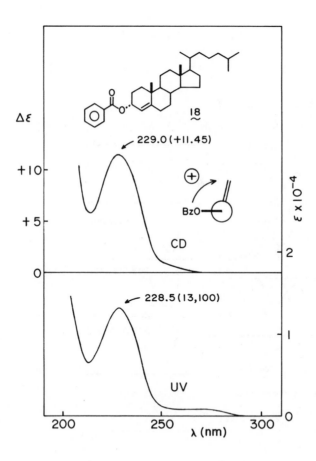

Figure 7–11.   *CD and UV spectra of cholest–4–en–3α–ol benzoate in ethanol.*

In the case of cholest-4-en-3α-ol benzoate (18), the CD spectrum exhibits a positive benzoate Cotton effect, the sign of which is in agreement with that of the exciton chirality between the benzoate and double bond groups (Figure 7-11). Thus, the α/β or R/S configuration of allylic alcohols is easily determinable in a straightforward manner from the CD spectra of benzoates.

In the case of cholest-4-ene-3β,6β-diol 6-benzoate (19), the C(4)-C(5) double bond is exocyclic to ring B to which the 6-benzoate is attached; the present exciton method also holds for such cases as shown in Figure 7-12.

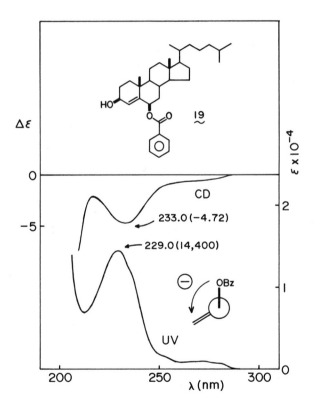

Figure 7-12. *CD and UV spectra of cholest-4-ene-3β,6β-diol 6-benzoate in ethanol.*

In the case of 5α-cholest-7-ene-3β,6α-diol 3-acetate 6-benzoate (<u>20</u>), the benzoate Cotton effect is positive in agreement with the positive exciton chirality (Figure 7-13). Unlike the usual cases discussed above, however, the ellipticity at shorter wavelengths is opposite in sign to the benzoate Cotton effect.   This is presumably due to the CD power of the strained C(7)-C(8) double bond.   A similar negative ellipticity at shorter wavelengths is also observed in the case of 5α-cholest-7-ene-3β,6β-diol 3-acetate 6-benzoate (<u>21</u>) (Figure 7-14).

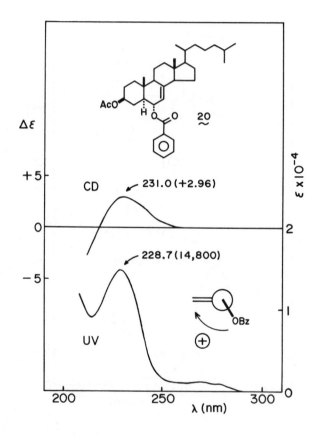

Figure 7-13. *CD and UV spectra of 5α-cholest-7-ene-3β,6α-diol 3-acetate 6-benzoate in ethanol.*

Other pertinent examples are listed in Table 7-9; all chiroptical data, including the case of monocyclic compounds, are in agreement with the exciton chirality between the benzoate and double bond chromophores.   The present exciton method for determining the absolute stereochemistry of <u>allylic</u> <u>alcohols</u> is based on theoretical grounds, and covers the empirical Mills' rule,[13] Brewster's rule,[14] and the benzoate sector rule.[15]

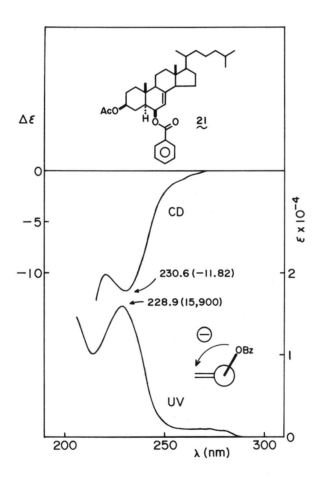

Figure 7-14.   *CD and UV spectra of 5α-cholest-7-ene-3β,6β-diol 3-acetate 6-benzoate in ethanol.*

Table 7-9.   **CD or ORD Data of Allylic Benzoates.**

Compound	Chirality	CD or ORD	Solvent
	⊕	a +258 (236/218)	M
	⊕	$\Delta[\emptyset]_D$ +253°	
	⊖	$\Delta\varepsilon$ −24.13 (224.5)	E
	⊖	$\Delta\varepsilon$ −8.93 (229.5)	E

$\Delta\varepsilon$ −7.84
(239.0)   E

$\Delta\varepsilon$ +6.44
(230.0)   E

$\Delta\varepsilon$ −11.81
(229.0)   E

$\Delta\varepsilon$ +7.26
(229.5)   E

$\Delta\varepsilon$ +4.88
(229.5)

E

$\Delta\varepsilon$ +4.32
(225)

M/D

$\Delta[\emptyset]_D$ +329°

$\Delta[\emptyset]_D$ −310°

References

1. M. Koreeda, N. Harada, and K. Nakanishi, J. Am. Chem. Soc. 96, 266 (1974).

2. V. Delaroff and R. Viennet, Bull. Soc.Chim. France, 1972, p. 277.

3. M. Koreeda, G. Weiss, and K. Nakanishi, J. Am. Chem. Soc. 95, 239 (1973).

4. M. A. Adams, K. Nakanishi, W. C. Still, E. V. Arnold, J. Clardy, and C. J. Persoons, J. Am. Chem. Soc. 101, 2495 (1979).

5. W. C. Still, J. Am. Chem. Soc. 101, 2493 (1979).

6. A. Moscowitz, E. Charney, U. Weiss, and H. Ziffer, J. Am. Chem. Soc. 83, 4661 (1961). U. Weiss, H. Ziffer, and E. Charney, Tetrahedron 21, 3105 (1965). E. Charney, H. Ziffer, and U. Weiss, Tetrahedron 21, 3121 (1965).

7. A. W. Burgstahler, R. C. Barkhurst, and J. K. Gawronski, in Modern Methods of Steroid Analysis ed. E. Heftmann (New York: Academic Press, 1973), Chapter 16.

8. N. Harada, unpublished data.

9. N. Harada, J. Iwabuchi, Y. Yokota, H. Uda, and K. Nakanishi. J. Am. Chem. Soc. 103, 5590 (1981).

10. K. Stich, G. Rotzler, and T. Reichstein, <u>Helv. Chim. Acta</u> <u>42</u>, 1480 (1959).

11. C. C. Levin and R. Hoffmann, <u>J. Am. Chem. Soc.</u> <u>94</u>, 3446 (1972), and references cited therein.

12. (a) A. Yogev, D. Amar, and Y. Mazur, <u>Chem. Commun.</u> 1967, p. 339.   (b) A. I. Scott and A. D. Wrixon, <u>Tetrahedron</u> <u>26</u>, 3695 (1970).

13. J. A. Mills, <u>J. Chem. Soc.</u> 1952, p. 4976.

14. J. H. Brewster, <u>Tetrahedron</u> <u>13</u>, 106 (1961).   See also J. H. Brewster, <u>J. Am. Chem. Soc.</u> <u>81</u>, 5475, 5483, 5493 (1959).

15. N. Harada, M. Ohashi, and K. Nakanishi, <u>J. Am. Chem. Soc.</u> <u>90</u>, 7349 (1968).   N. Harada and K. Nakanishi, <u>J. Am. Chem. Soc.</u> <u>90</u>, 7351 (1968).

16. Applications of the allylic benzoate method: (a) eleven membered allylic alcohol of sesterterpene, Y. Naya, K. Yoshihara, T. Iwashita, H. Komura, K. Nakanishi, and Y. Hata, <u>J. Am. Chem. Soc.</u>, <u>103</u>, 7009 (1981); (b) synthetic five membered allylic alcohol, W. H. Rastetter, J. Adams, and J. Bordner, <u>Tetrahedron Lett.</u>, <u>23</u>, 1319 (1982).

VIII. CHIRAL EXCITON INTERACTION BETWEEN OTHER BENZENOIDS AND/OR HETEROAROMATICS

8-1. Exciton Interaction in Bisindole Alkaloids

Dimeric bisindole alkaloids have two similar chromophores, the indole and m-methoxyaniline chromophores chirally disposed. The spatial interaction should generate CD activity.

Kutney et al.,[1] rationalized the circular dichroism of a number of bisindole alkaloids with antitumor activity and developed an empirical chiroptical method to determine the configuration at C-18' that links the indole and

Table 8-1.

224 (+30)

207 (-69)

CD (1)

MeOH

229 (+19)	CD (1)
205 (-37)	X ray (2)
MeOH	

221 (-57)	CD (1)
210 (+44)	X ray (3)
MeOH	

dihydroindole units. The UV spectrum of the indole moiety has a strong $\pi \rightarrow \pi*$ ^1B_b transition around 225 nm (ε about 19,000), while the indoline chromophore has its benzenoid ^1B transition around 217 nm (ε about 32,000). Since these two intense transitions are close, both in wavelength position and in space, exciton split CD interactions between these two transitions were expected. In fact, the natural bisindoles 1 and 2 possessing 18'α-indoline unit exhibit typical exciton split Cotton effects of positive first and negative second signs in the corresponding wavelength region. On the other hand, synthetic 3 with the unnatural configuration at 18'-position, i.e., 18'β-indoline unit, shows negative first and positive second Cotton effects (Table 8-1). These phenomena were also observed in other bisindole alkaloids. It is hence clear that the observed split Cotton effects are mainly due to the exciton coupling between indole and indoline chromophores and that the sign of the Cotton effects reflects the configuration at C-18'. The present CD studies provide a simple method for evaluating the chirality of compounds at this chiral center which is crucial from the viewpoint of bioactivity.

In order to determine the absolute configuration of bisindoles in a non-empirical manner, however, by applying the exciton chirality method, it is necessary to evalute the polarization properties of the electric transition moments involved. This is difficult because of low-symmetrical structure of both chromophores and remains to be solved.

8-2. Exciton CD Spectra of N-(5-Bromosalicylidene) Derivatives of Diamines

Smith and coworkers[4] reported the CD spectra of N-(5-bromosalicylidene) derivatives of diamine systems, which exhibit several split Cotton effects below 264

Table 8-2.

	415 (1,300)	415 (−2)
	328 (6,900)	337 (−17)
	280 (3,100)	
		264 (−30)
	247 (18,000)	*247 (+33)sh*
		233 (+130) CD (4)
	222 (62,000)	*217 (−67)*
4	MeOH	MeOH

414 (1,400)	415 (−1)	
328 (6,800)	336 (−7)	
254 (19,000)	*263 (−15)*	
247 (19,000)	*247 (+11)*	CD (4)
222 (59,000)	*232 (+57)*	
	216 (−51)	

<u>5</u>   MeOH   MeOH

nm, as exemplified in Table 8-2. These Cotton effects were interpreted on the basis of the exciton coupling mechanism; the Cotton effects of medium intensity at 264 and 247 nm are of negative chirality, while those at 233 and 217 nm are of positive chirality. However, the obtained chiroptical data did not provide direct configurational information because the chromophore is not symmetrical, thus leading to an uncertainty in the polarization properties. Therefore, they inversely utilized these Cotton effects to assist in determining the transition moment directions of the chromophore.

8-3.   Exciton Circular Dichroism of Some Nucleotides

Similar to the dimeric compounds discussed above, dinucleotides and related polymers also exhibit exciton split CD Cotton effects around 260 nm. By employing such exciton split Cotton effects, configurational and conformational studies of various di-, obligo-, and poly-nucleotides have been carried out by Tinoco, Cantor, and others. In spite of the significance of such molecules, only a few examples are discussed in the following because the biochemical theme is beyond the scope of this book.

Table 8-3.

271 (+6.2)
252 (-7.6)
220 (+6.1)   CD (5)

0.01 M Na_3PO_4 buffer

0.1 M $NaClO_4$

pH 7.2, 26°C

A typical example of a dinucleotide is adenylyl-(3'-5')-adenosine, ApA,[5] the CD spectrum of which shows positive first and negative second Cotton effects at 271 and 252 nm, respectively (Table 8-3); the π→π* transitions of the two adenyl chromophores at 260 nm are coupled to each other to give split Cotton effects of positive chirality. Contrary to the cases discussed in previous chapters, however, it is rather difficult to deduce the conformation solely from these data because of conformational flexibility and the low symmetrical structure of the adenine chromophore.

Urry and coworkers[6] studied the biomolecular conformations by applying the CD spectroscopic method to various bis-chromophoric systems. For example, adenosine-5'-mononicotinate (7) exhibits typical exciton split Cotton effects as shown in Figure 8-1.[6,7] The π→π* transition of adenosine at 259 nm couples with the less intense transition of nicotinate at 264 nm to give negative first and positive second Cotton effects, which indicates a stacked conformation at low temperatures. When the temperature is increased, the CD amplitude decreases (Figure 8-1); this was interpreted as being caused by destacking at high temperatures.

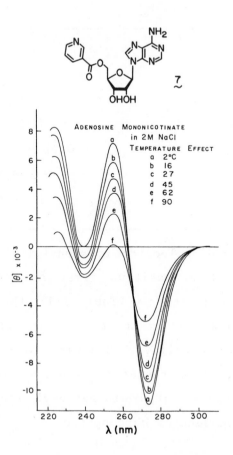

Figure 8-1.   *CD spectra of adenosine-5'-mononicotinate in aq. 2M NaCl and its temperature dependence. [Reprinted from reference 6.]*

The appearance of a typical split CD was utilized in elucidating the bis-chromophoric structure of an orotidine 5'-monophosphate decarboxylase inhibitor **8**.[8] In this case, although only 20 µg of the sample was available, the

Table 8-4.

	261 (35,000)	274 (+34)	CD (8)
		258 (-17)	
	H_2O, pH 7	H_2O, pH 7	

_8_

	260 (13,000)	290 (-0.20)	CD (8)
	228 (4,800)sh		
	H_2O, pH 7	H_2O, pH 7	

_9_

| | 262 (22,000) | 262 (+0.62) | CD (8) |
| | H_2O, pH 7 | H_2O, pH 7 | |

_10_

intense split Cotton effects indicated that two aromatic chromophores were present in the same molecule: the CD spectrum of <u>8</u> exhibits two split Cotton effects around 260 nm; i.e., λ_{ext} 274 nm, $\Delta\varepsilon$ +34 and 258 nm, $\Delta\varepsilon$ −17 (in water).   On the other hand, two other mono-chromophoric inhibitors <u>9</u> and <u>10</u> showed only a single and weak CD extremum (Table 8-4). This led to structure <u>8</u>, which was subsequently proven by synthesis with barbituric acid and ribose 5-phosphate.

8-4.   pK$_a$ Determination by the Exciton Coupling CD Spectroscopy

The exciton coupling CD spectroscopy also provides a convenient micromethod for determining pK$_a$ values of complex compounds on a microgram scale.[9] The method is useful, for example, for determining the point of attachment of carcinogenic hydrocarbons to nucleic acid bases in the adducts formed between these two moieties; this cannot be determined from ^{1}H-NMR because the bases

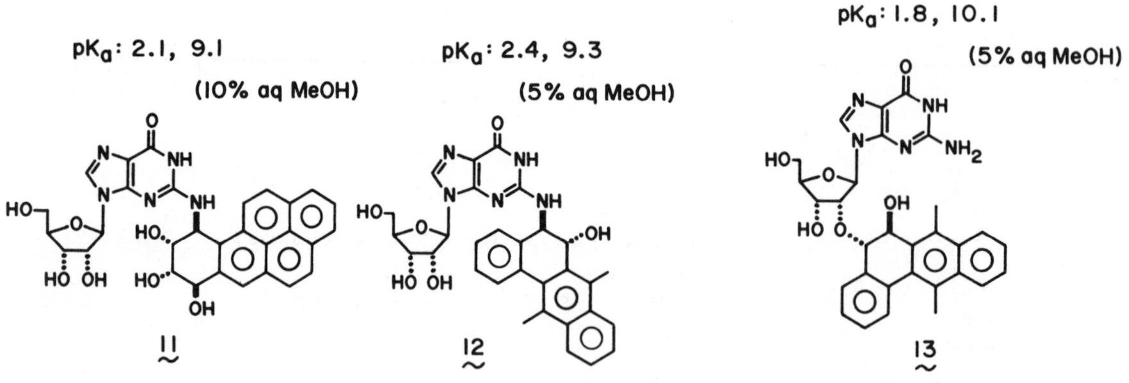

Figure 8-2. pK$_a$ values of guanosine-polyaromatic hydrocarbon adducts.

carry only a few protons and furthermore, they can exist in several tautomeric modifications.   In the case of guanosine-benzo[a]pyrene adduct (11)[10], the site of linkage on the base unit was determined as follows.   The number and values of pK$_a$ are dependent on the substitution pattern of guanosine derivatives as shown in Figure 8-3.   The dissociation constants can neither be measured by titration, owing to the limited quantity of sample available, nor by UV spectroscopy, owing to the domination of spectra by the strong non-dissociating polyaromatic hydrocarbon chromophore (Figure 8-4).

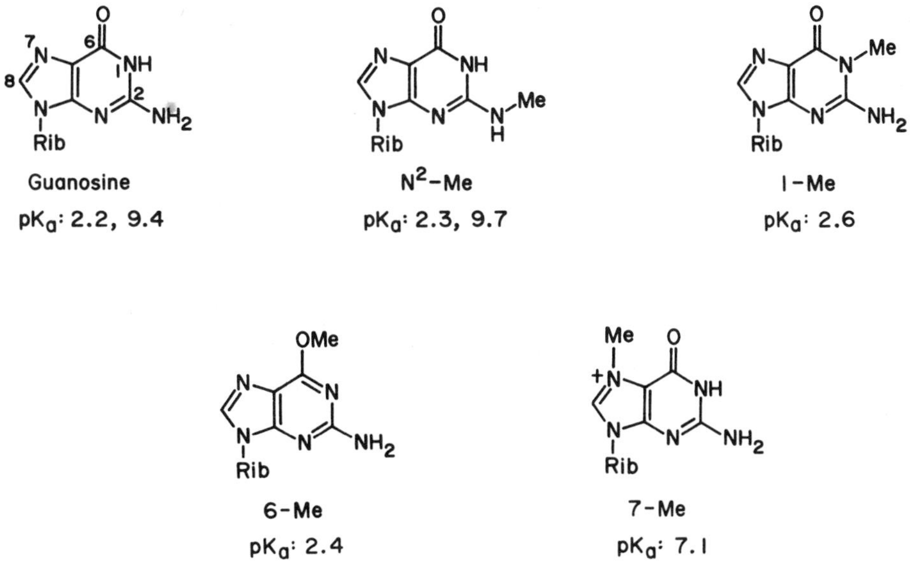

Guanosine
pK$_a$: 2.2, 9.4

N^2-Me
pK$_a$: 2.3, 9.7

I-Me
pK$_a$: 2.6

6-Me
pK$_a$: 2.4

7-Me
pK$_a$: 7.1

Figure 8-3. pK$_a$ *values of guanosine and methylated guanosines. [Values are taken from B. Singer,* Progr. Nucleic Acid Res. 15, *219 (1975); and N. K. Kochetkov and E. I. Budovskii,* Organic Chemistry of Nucleic Acids, *trans. B. Haigh (New York: Plenum Press, 1972)].*

Conveniently, however, the CD spectrum is complex and consists of several intense extrema because of the exciton interaction between the hydrocarbon and guanine chromophores (Figure 8-4).   Thus, although the CD curve has extrema at wavelengths corresponding to the hydrocarbon absorption maxima, the extrema

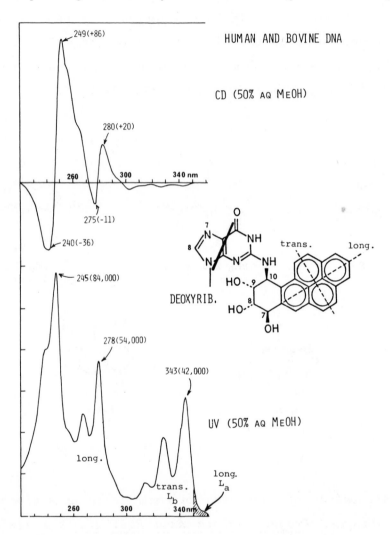

Figure 8-4. CD and UV spectra of guanosine-benzo[*a*]pyrene tetrahydrodiol epoxide adduct in 50% aq MeOH. [Reprinted from reference 10.]

are the result of spatial interactions and therefore are subject to changes in
the guanosine charge.

A plot of CD against pH leads to measurements of pK$_a$ values as shown in
Figure 8-5.   Apparent dissociation constants of 2.1 and 9.1 (in 10% aq MeOH)
were obtained when the changes of $\Delta\varepsilon$ at the 250 nm extrema ($\Delta\varepsilon$ + 72 at pH 7.1)
were plotted against the pH.   Since two pK$_a$ values were obtained in the
range of pH 1 to 11, the hydrocarbon part had to be attached to C-8, 2-NH$_2$
or the ribose.   The present method was crucial in determining the linkage
sites of guanosine-7,12-dimethylbenz[a]anthracene adducts (12) and (13).[9,11]

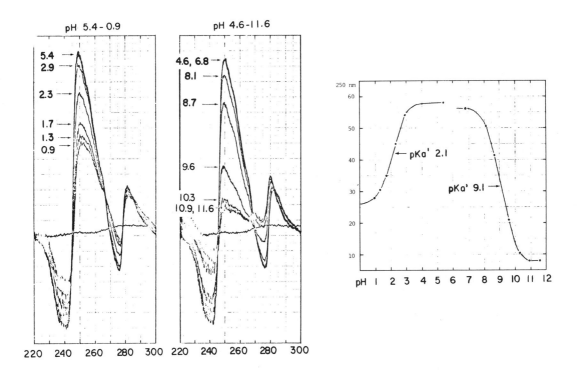

Figure 8-5. *Change in CD spectrum of adduct (11) with pH in 10% aq MeOH, and
plot of CD value at 250 nm against pH. [Reprinted from reference 10.]*

8-5. Alkaloids with Two o-Dimethoxybenzene Chromophores

In alkaloids containing two o-dimethoxybenzene chromophores, there is some controversy as to whether the observed CD Cotton effects are due to the coupled mechanism or not.

The SCF-CI molecular orbital calculation results[12] of the o-dimethoxybenzene chromophore indicate that the weak absorption band around 280 nm originates from the ^1L_b transition of benzene and is polarized along the x-axis, while the ^1L_a transition around 230 nm is polarized along the y-axis (Table 8-5).

The intense absorption below 200 nm originates from the ^1B transition of benzene. As listed in Table 8-5, the small splitting between ^1B_b and

Table 8-5. Observed and Calculated UV Spectra of o-Dimethoxybenzene[12]

Observed (hexane)			Calculated (SCF-CI)		
λ_{max} (nm), (ε)	$D \times 10^{36}$		λ_{max} (nm)	$D \times 10^{36}$	Polarization
278.0 (2,400)	1.82	^1L_b	278.6	1.55	x-axis
227.8 (7,200)	6.30	^1L_a	227.4	5.92	y-axis
198.3 (48,300)	31.50	^1B	194.9	46.76	x-axis
			193.1	34.41	y-axis

^1B_a states suggest that the observed transition is degenerate. Therefore, polarization of the transition cannot be defined. Accordingly, the ^1B transition is not suited, in general, for the exciton coupling method.

Mason and coworkers[13] determined the absolute configuration of (−)-argemonine (14) by analysing the CD spectrum in terms of the coupled oscillator theory. The CD spectrum of argemonine exhibits weak negative first and positive second Cotton effects in the ^1L_b transition region; a negative exciton chirality was therefore assigned to (−)-argemonine, which led to the

Table 8-6.

	289 (8,200)	*297 (−1.3)*	
		275 (+5.1)	
	226 (14,000)	234 (−29)	CD (13)
		208 (+)	
			Correl. (14)
		EtOH/isopentane/	
		ether	
		2 : 5 : 5	
14		at 80° K	

($1\underline{S},5\underline{S}$) absolute stereochemistry (Figure 8-6).   The numerical calculation based on the coupled oscillator theory supported the conclusion.   The deduced absolute configuration was further verified by chemical correlation with ($\underline{S}$)-(+)-aspartic acid.[14]

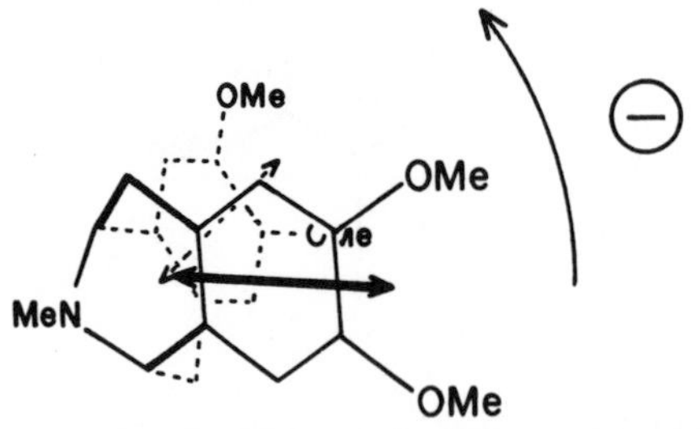

Figure 8-6.   *Negative exciton chirality between   the   two ^1L_b transitions   in the system of (-)-argemonine.   The transitions are polarized along the short axis of $\underline{o}$-dimethoxybenzene chromophore.   [Adapted from reference 13.]*

However, Snatzke and coworkers[15] have criticized the coupled oscil-lator treatment.   Namely, the bisignate CD was found only in the ^1L_b tran-sition region, but the pattern is highly unsymmetric.   Furthermore, norarge-monine (15) exhibits no first negative CD band.   Rather, they interpreted the

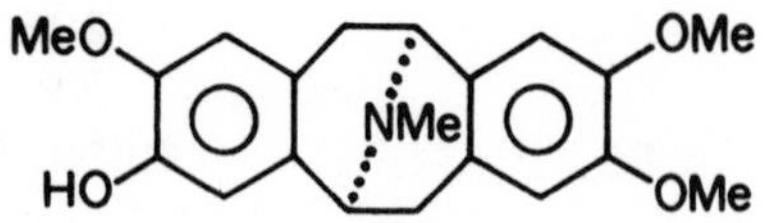

15

CD of argemonine and its analogues as a summation of the Cotton effect of each half. Instead of the coupled oscillator method, they applied the empirical sector rule of the benzene chromophores.

Table 8-7.

291 (7,400)	298 (+14.9)	
	280 (-4.0)	
239 (8,100)	240 (+9.0)	CD (16)
201 (81,300)		
EtOH	EtOH	

291 (4,800)	297 (+7.3)	
	280 (-2.9)	CD (16)
239 (5,800) sh	240 (+3.8)	
EtOH	EtOH	

By applying the coupled oscillator method, absolute configurations of spirobenzylisoquinoline alkaloids, (+)-chorobiline and dihydrofumariline, were deduced;[16] the positive first and negative second signs of the bisignate CD Cotton effects in the ^1L_b region led to the absolute configurations depicted in Table 8-7.

In the case of (-)-amurensine ($\underline{18}$),[17] the CD spectrum clearly shows bisignate CD Cotton effects in the ^1L_b region, and the absolute configuration depicted in Table 8-8 was deduced by application of the exciton chirality method (Figure 8-7). This assignment indicates that argemonine $\underline{14}$ and amurensine $\underline{18}$ have the same absolute configuration; the two o-dimethoxybenzene chromophores constitute a left-handed screwness.

It is tempting to analyse the complex CD spectrum of amurensine $\underline{18}$ as follows. In addition to the bisignate ^1L_b Cotton effects, one notices that the UV shoulder at 233 nm (^1L_a) is also associated with negative

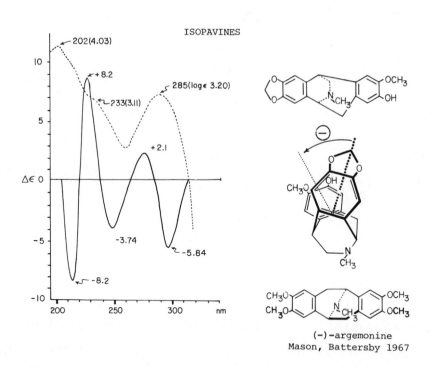

ISOPAVINES

(-)-argemonine
Mason, Battersby 1967

Figure 8-7. CD (solid line) and UV (dotted line) spectra of (-)-amurensine in ethanol. [Adapted from reference 17].

Table 8-8.

	285 (1,600)	(−5.8)	
		(+2.1)	
	233 (1,300)sh	(−3.7)	CD (17)
		(+8.2)	
	202 (10,700)	215 (−8.2)	

18

(−3.74) and positive (+8.2) Cotton effects (Figure 8-7). The signs of the split ^1L_a CD curve are in agreement with the result derived from analysis of the ^1L_b region.

As indicated in Table 8-5, the ^1L_b transition of o-dimethoxybenzene originates from the theoretically forbidden band of benzene. Accordingly, the transition is weak (ε 2,400), and the exciton split Cotton effects, if any, would also be weak. Therefore, it is difficult to discriminate whether the observed Cotton effects are due to the exciton coupling or a simple summation. Thus, although the ^1L_b transition of the o-dimethoxybenzene chromophore has been used in amurensine, the results derived are less reliable in comparison with the cases of dibenzoates.

The alkylated benzene chromophores are even less suited. Benzene itself has three short axes of ^1L_b transition, and the substituent effect of an alkyl group is too weak to specify one of these axes. Furthermore, the strength of the ^1L_b transition is weak, and consequently the coupling is weak.

In conclusion, in determining absolute configurations in a reliable manner, it is necessary to choose proper electronic transitions of appropriate chromophores which satisfy the requirements discussed in Chapter 3. It should be remembered that there are only two possible conclusions in absolute configurational studies: the conclusion is either correct or wrong.

References

1. J. P. Kutney, D. E. Gregonis, R. Imhof, I. Itoh, E. Jahngen, A. I. Scott, and W. K. Chan, J. Am. Chem. Soc. 97, 5013 (1975).

2. J. W. Moncrief and W. N. Lipscomb, J. Am. Chem. Soc. 87, 4963 (1965).

3. J. P. Kutney, J. Cook, K. Fuji, A. M. Treasurywala, J. Clardy, J. Fayos, and H. Wright, Heterocycles 3, 205 (1975).

4. H. E. Smith, J. R. Neergaard, E. P. Burrows, and F. -M. Chen, J. Am. Chem. Soc. 96, 2908 (1974).

5. M. M. Warshow and C. R. Cantor, Biopolymers 9, 1079 (1970).

6. D. W. Urry in Spectroscopic Approaches to Biomolecular Conformation, ed. D. W. Urry, (Chicago: American Medical Assoc. 1970), Chapter 3.

7. D. W. Miles and D. W. Urry, J. Phys. Chem. 71, 4448 (1967).

8. H. Komura, K. Nakanishi, B. W. Potvin, H. J. Stern, and R. S. Krooth, J. Am. Chem. Soc. 102, 1208 (1980).

9. H. Kasai, K. Nakanishi, and S. Traiman, J. Chem. Soc., Chem. Commun. 1978, p. 798.

10. K. Nakanishi, H. Kasai, H. Cho, R. G. Harvey, A. M. Jeffrey, K. W. Jennette, and I. B. Weinstein, J. Am. Chem. Soc. 99, 258 (1977).

11. H. Kasai, K. Nakanishi, K. Frenkel, and D. Grunberger, J. Am. Chem. Soc.
     99, 8500 (1977).

12. N. Harada, unpublished data.

13. (a) S. F. Mason, G. W. Vane, and J. S. Whitehurst, Tetrahedron 23, 4087
     (1967); (b) S. F. Mason, K. Schofield, R. J. Wells, J. S. Whitehurst, and
     G. W. Vane, Tetrahedron Lett. 1967, p. 137.

14. (a) A. R. Battersby and A. C. Barker, Tetrahedron Lett. 1967, p. 135; (b)
     A. C. Barker and A. R. Battersby, J. Chem. Soc. (c) 1967, p. 1317.

15. G. Snatzke, M. Kajtar, and F. Snatzke, in Fundamental Aspects and Recent
     Developments in Optical Rotatory Dispersion and Circular Dichroism, ed.
     F. Ciardelli and P. Salvadori, (London: Heyden 1973), Chapter 3-4.

16. M. Shamma, J. L. Moniot, R. H. F. Manske, W. K. Chan, and K. Nakanishi,
     J. Chem. Soc., Chem. Commun. 1972, p. 310.

17. M. Shamma, J. L. Moniot, W. K. Chan, and K. Nakanishi, Tetrahedron Lett.
     1971, p. 3425.

18. Chiral molecular association of natural flower pigments: T. Hoshino, U.
     Matsumoto, N. Harada, and T. Goto, Tetrahedron Lett. 22, 3621 (1981).

IX. EMPIRICAL METHODS EMPLOYING ORGANOMETALLIC REAGENTS

9-1. A Method for Determining Absolute Configuration of Vicinal Glycols and Amino Alcohols with Pr(dpm)$_3$

Addition of optically active α-glycols or α-amino alcohols to a solution of metal compounds results in induced CD Cotton effects in ultraviolet and/or metallic d-d transition regions. Correlation of the sign of induced CD Cotton effects to absolute configuration of glycols or amino alcohols provides empirical methods for determining absolute configuration of unknown compounds.

A typical case is the cuprammonium method,[1] which correlates the sign of metallic d-d Cotton effect with the chiralities of glycols and amino alcohols; a positive chirality is associated with a positive Cotton effect at 580-600 nm when the compound is measured in aqueous cuprammonium solution. The method is an extension of the classical Reeves' method[2] used extensively in carbohydrates.

Recently, it was found that the NMR shift reagent tris(dipivalomethanato)praseodymium, Pr(dpm)$_3$ (1), exhibits bisignate induced CD Cotton effects centered around 300 nm, when the organometallic reagent and vicinal glycols or amino alcohols were mixed in CCl$_4$ or hexane.[3,4] For example, the CD spec-

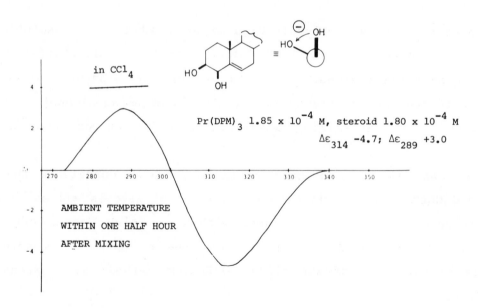

trum of a mixture of Pr(dpm)$_3$ and cholest-5-ene-3β,4β-diol (2) (ca. 1:1) shows negative first and positive second Cotton effects, λ_{ext} 314 nm (Δε −4.7) and 289 (+3.0), as illustrated in Figure 9-1 where the Δε values are based on the concentration of Pr(dpm)$_3$.   In this case, the negative sign of the first

Pr(DPM)$_3$ 1.85 x 10^{-4} M, steroid 1.80 x 10^{-4} M

Δε$_{314}$ −4.7; Δε$_{289}$ +3.0

Figure 9-1.   CD spectrum of a mixture of Pr(dpm)$_3$ and cholest-5-ene-3β,4β-diol in CCl$_4$.   [Adapted from reference 4.]

Cotton effect agrees with the negative chirality of α-glycol moiety. The following empirical method was induced from the accumulated data of a number of glycols of known absolute configuration:[4] a positive chirality is associated with positive first Cotton effect around 303-315 nm and vice versa.

In the case of acyclic systems, chirality is defined as follows: the conformer depicted in Figure 9-2 where bulkier groups L are pseudo-equatorial is assumed to be involved in complex formation. Then, a negative chirality is defined for the counter-clockwise screwness between two hydroxyl groups shown in Figure 9-2.

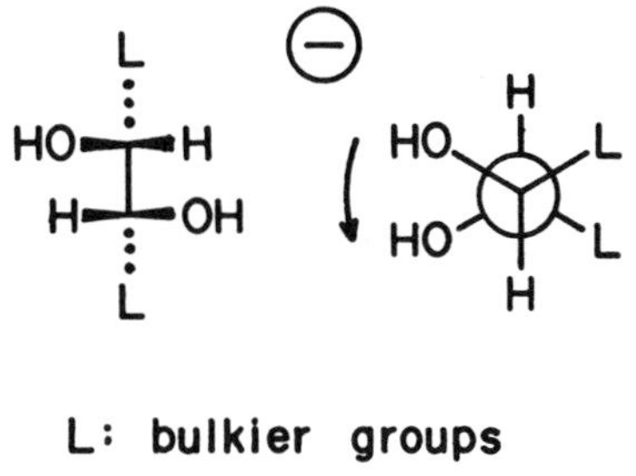

L: bulkier groups

Figure 9-2. Conformer to be considered in the present organometallic methods, where bulkier groups L adopt pseudo-equatorial position. [Adapted from reference 4.]

Absolute configurations of the following natural products have been determined by the present empirical method. Epoxide cleavage of the natural insect juvenile hormone yielded the glycol **3**; application of this method, coupled with the measurements of several other glycols of established configurations, determined the absolute configuration of the epoxide as 10R,11S;[5] independent chemical correlation studies[6] led to the same conclusion.

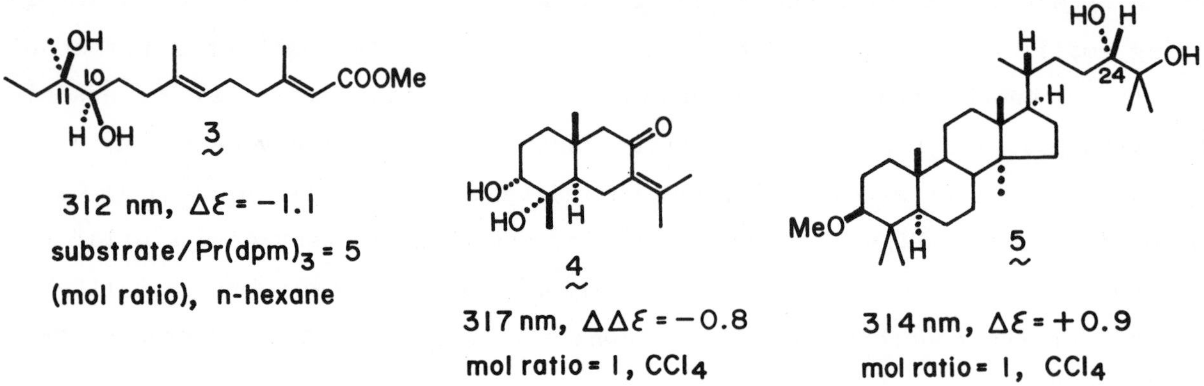
OH
10
11
H OH
COOMe
3
312 nm, ΔE = −1.1
substrate/Pr(dpm)3 = 5
(mol ratio), n-hexane
O
HO
HO
H
4
317 nm, ΔΔE = −0.8
mol ratio = 1, CCl4
HO H
H
OH
24
H
MeO
H
5
314 nm, ΔE = +0.9
mol ratio = 1, CCl4

Figure 9-3. CD Cotton effects of mixtures of $Pr(dpm)_3$ and vicinal glycols: juvenile hormone derivative, sesquiterpene cuauhtemone, and triterpene.

Compound 4 is the sesquiterpene cuauhtemone, where flexibility of the enone group prevented application of the rules proposed for determining the enone helicity. A differential CD curve before and after the addition of $Pr(dpm)_3$ to a CCl_4 solution of the compound showed a negative band at 317 nm ($\Delta\Delta\varepsilon$ −0.8), and this led to the absolute configuration depicted.[7]

Kutney and coworkers[8] deduced the absolute configuration of the C-24/C-25 glycol moiety in triterpene 5 by means of the present method. The positive Cotton effect led to a (24S) configuration (Figure 9-3). This is in line with the results of the next steroidal compounds.

Partridge, et al.,[9] have shown that another lanthanide shift reagent, tris(6,6,7,7,8,8,8-heptafluoro-2,2-dimethyl-3,5-octanedionato)europium, $Eu(fod)_3$, (6) gives stronger induced CD's. They have used this reagent in deducing the absolute configurations at C-24/C-25 of the steroidal side chain

in   24,25-dihydroxycholesterols,   (7)   and   (8),   after   establishing   the
configuration of a derivative by X-ray crystallography (Figure 9-4).

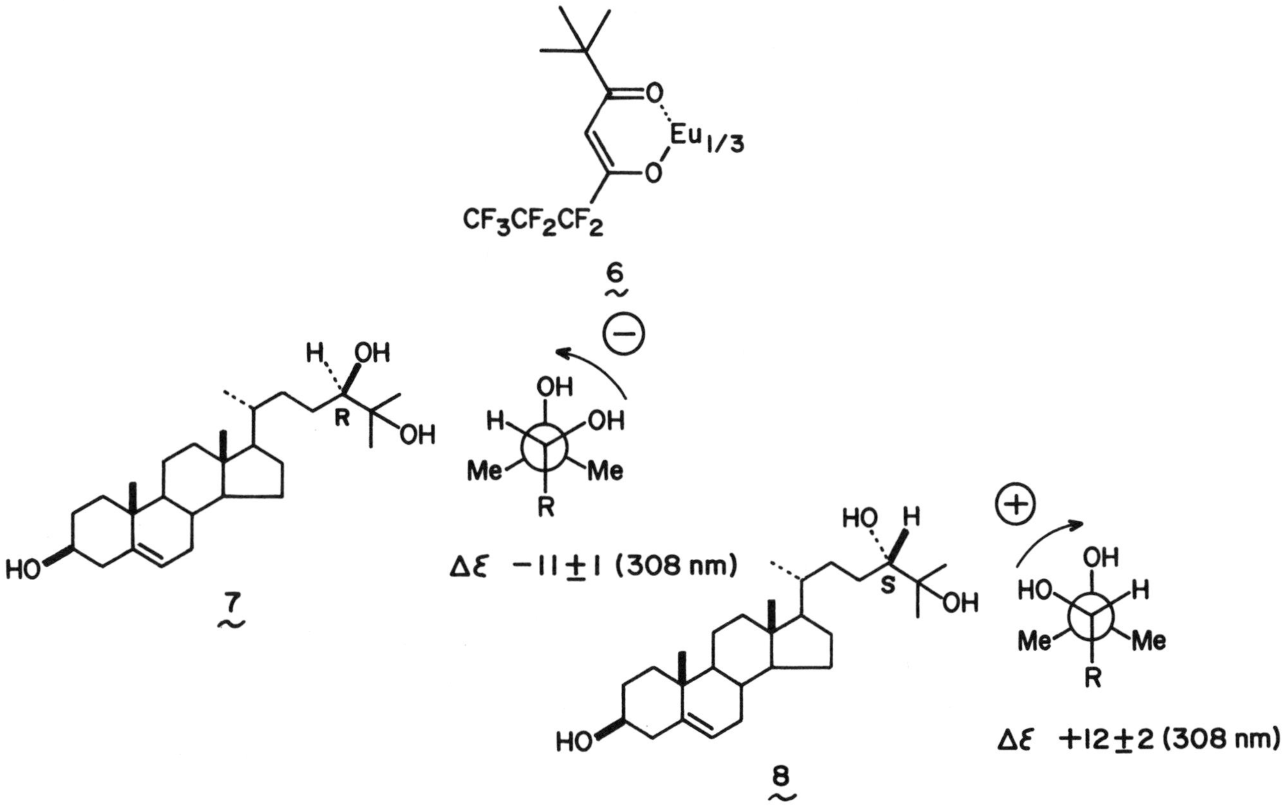

Figure 9-4. CD Cotton effects of mixtures of Eu(fod)$_3$ and 24,25-dihydroxy-cholesterols.

9-2.   A Method for Acyclic Glycols and Amino Alcohols with
Ni(acac)$_2$

Interaction of nickel bis(acetylacetonate), Ni(acac)$_2$, with various glycols
in CCl$_4$ or t-BuOH/CCl$_4$ was studied by UV and CD spectroscopic methods,[10-12] and the following generalities were induced: a negative chirality

is associated with a weak negative d–d Cotton effect around 630 nm, a positive extremum around 315 nm, and a negative peak around 293 nm (Figure 9–5).

For instance, a solution of Ni(acac)$_2$ and (2$\underline{R}$,3$\underline{R}$)–butane–2,3–diol ($\underline{9}$) (mol ratio 1:40) in CCl$_4$ exhibits CD and UV spectra shown in Figure 9–5. In this case, the conformer with the two bulky substituents (methyl groups) in the pseudo–equatorial position was assumed to be involved in complexation and this conformer with a negative chirality between two hydroxyl groups leads to

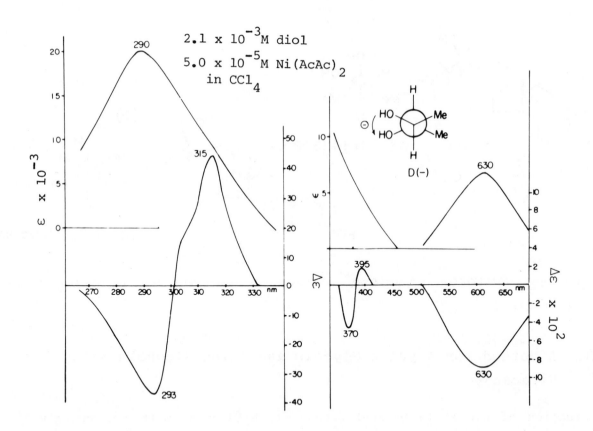

Figure 9–5. _UV and CD spectra of a mixture of Ni(acac)$_2$ and (2$\underline{R}$,3$\underline{R}$)–_
butane–2,3–diol in CCl$_4$. [Adapted from reference 12.]

the signs of Cotton effects depicted in Figure 9-5. Similar Cotton effects were observed for various acyclic glycols. However, the present method gave opposite results when applied to the cyclic glycol system present in cholest-5-ene-3β,4β-diol (2).[10] It can be applied to acyclic systems only.

9-3. Applicability of the Organometallic Methods

The origin of the bisignate CD Cotton effects induced by the organometallic reagents is not clear, although the coupled interaction between $\pi \rightarrow \pi^*$ transitions of the β-diketo ligands, dipivalomethanato or acetylacetonate, was proposed as one possible origin.[12] Accordingly, the present methods are based on empirical generalities. Furthermore, these empirical methods have led to configurations opposite to the actual case.[4,12,13] Hence, when applying these methods, it is essential to choose reference compounds with closely related structural moieties and of known absolute configurations.

References

1. S. T. K. Bukhari, R. D. Wrixon, A. I. Scott, and A. D. Wrixon, Chem. Commun. 1968, p. 1580; Tetrahedron 26, 3653 (1970).

2. R. E. Reeves, Adv. Carbohydr. Chem. 6, 107 (1971).

3. K. Nakanishi and J. Dillon, J. Am. Chem. Soc. 93, 4058 (1971).

4. J. Dillon and K. Nakanishi, J. Am. Chem. Soc. 97, 5417 (1975).

5. K. Nakanishi, D. A. Schooley, M. Koreeda, and J. Dillon, J. Chem. Soc., Chem. Commun. 1971, p. 1235.

6. D. J. Faulkner and M. R. Petersen, J. Am. Chem. Soc. 93, 3766 (1971).

7. K. Nakanishi, R. Crouch, I. Miura, X. Dominguez, A. Zamudio, and R. Villarreal, J. Am. Chem. Soc. 96, 609 (1974).

8. J. P. Kutney, G. Eigendorf, R. B. Swingle, G. E. Knowles, J. W. Rowe, and B. A. Nagasampagi, Tetrahedron Lett. 1973, p. 3115.

9. J. J. Partridge, V. Toome, and M. R. Uskokovic, J. Am. Chem. Soc. 98, 3739 (1976).

10. J. Dillon and K. Nakanishi, J. Am. Chem. Soc. 96, 4057 (1974).

11. J. Dillon and K. Nakanishi, J. Am. Chem. Soc. 96, 4059 (1974).

12.   J. Dillon and K. Nakanishi, J. Am. Chem. Soc. 97, 5409 (1975).

13.   See Saperin B in section 3-3-A: C. W. Lyons and D. R. Taylor, J. Chem. Soc., Chem. Commun., 1976, p. 647.

X. QUANTUM MECHANICAL THEORY OF THE EXCITON CHIRALITY METHOD

10-1. Rotational Strength of CD Cotton Effects

Rotational strength R, a theoretical parameter representing the sign and strength of a CD Cotton effect, is formulated in equation 10.1 and is experimentally obtainable from the observed CD spectra:[1-4]

$$R = 2.296 \times 10^{-39} \int_0^\infty \Delta\epsilon(\sigma)/\sigma \; d\sigma \qquad \text{cgs unit} \qquad (10.1)$$

$$= (2.296 \times 10^{-39}/\sigma_0) \int_0^\infty \Delta\epsilon(\sigma) \; d\sigma$$

where σ is wavenumber and σ_0 is the wavenumber of the extremum of the Cotton effect. The value can be evaluated as follows: for example, in the case of the $n\rightarrow\pi^*$ Cotton effect (λ_{ext} 295.5 nm, $\Delta\epsilon$ +1.540) of (+)-camphor, the CD spectrum is plotted against wavenumber σ as shown in Figure 10-1. The integration term of equation 10.1 corresponds to the peak area of the Cotton effect. Thus, rotational strength R can be calculated by evaluating the band area. In the present case, numerical calculation gives R= +4.175 × 10^{-40} cgs unit.[5]

On the other hand, the rotational strength is theoretically expressed by equation 10.2, which was derived by Rosenfeld in the early days of the development of quantum theory:[6]

$$R = Im\ \{<0|\vec{\mu}|a>\cdot<a|\vec{M}|0>\} \tag{10.2}$$

where R is rotational strength for the excitation $0 \rightarrow a$, Im denotes the imaginary part of the term in brackets; $<\ >$ denotes the integration over configuration space, and $\vec{\mu}$ and $\vec{M}$ are operators of electric and magnetic moment vectors, respectively.   The dot • stands for scalar product of two vectors, and 0 and a

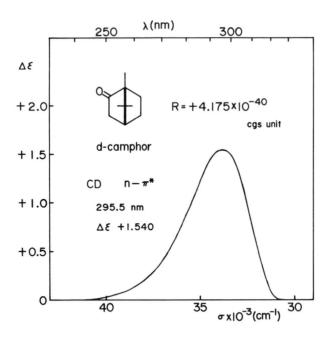

Figure 10-1. CD spectrum of (+)-camphor in methanol.

are wave functions of ground and excited states, respectively. Thus, rotational strength R is equal to the imaginary part of the scalar product of electric and magnetic transition moments. Since wave functions are, in general, real numbers, the electric and magnetic transition moments are real and imaginary, respectively.

Next, if a Gaussian distribution is approximated for a CD Cotton effect, the CD curve is formulated as

$$\Delta\epsilon(\sigma) = \Delta\epsilon_{max} \exp\{-((\sigma-\sigma_0)/\Delta\sigma)^2\} \tag{10.3}$$

where $\Delta\epsilon_{max}$ is the maximum value of the Cotton effect, and $\Delta\sigma$ is the standard deviation of the Gaussian distribution. Substitution of equation 10.3 into equation 10.1 and integration gives

$$R = 2.296 \times 10^{-39} \sqrt{\pi} \, \Delta\epsilon_{max} \, \Delta\sigma/\sigma_0 \tag{10.4}$$

From equations 10.4 and 10.3, the calculated CD curve is derived as follows:

$$\Delta\epsilon(\sigma) = (\sigma_0/(2.296 \times 10^{-39}\sqrt{\pi} \, \Delta\sigma)) \, R \, \exp\{-((\sigma-\sigma_0)/\Delta\sigma)^2\} \tag{10.5}$$

where σ_0, the excitation wavenumber of transition $0{\to}a$, and $\Delta\sigma$, the standard deviation of the Gaussian distribution, can be evaluated from the corresponding observed UV spectrum. Thus, if rotational strength R is theoretically calculable by equation 10.2, CD spectrum is reproducible by theoretical calculations.

10-2.   Dipole Strength and UV Spectra

In analogy with the case of rotational strength, dipole strength D representing the transition probability of the UV absorption band is expressed by
equation 10.6 and can be estimated from the observed UV spectra:[1-4]

$$D = 9.184 \times 10^{-39} \int_0^\infty \varepsilon(\sigma)/\sigma \, d\sigma \qquad \text{cgs unit} \qquad (10.6)$$

$$= \{<0|\vec{\mu}|a>\}^2 = \vec{\mu}^2 = (\vec{er})^2 \qquad (10.7)$$

where $\vec{\mu}$ is electric transition moment and $\vec{r}$ is transition length.

For example, in the case of the intramolecular charge transfer transition
of cholesterol p-chlorobenzoate,[1] integration over the absorption area from
263.0 nm to 214.8 nm gives the dipole strength value of 2.02×10^{-35} cgs unit
(Figure 10-2). From this value, the transiton length r of 0.9917 Å was obtained
by employing equation 10.7, where elementary charge $e = 4.803 \times 10^{-10}$ (esu) was
adopted.   Thus, transition length r representing the length of electric
transition dipole moment is experimentally obtainable from observed UV
spectra.

The electric transition moment is also theoretically calculable as indicated in equation 10.7, if wave functions of ground and excited states are
obtainable.   However, the theoretically calculated value is in general too
large to account for the strength of the observed UV spectra.   Therefore,
empirical values obtained from actual UV spectra are better than theoretical
values, and have been adopted hereafter.

As illustrated in Figure 10-2, the experimental UV band plotted against wavelength is asymmetric; namely, the longer wavelength side is steeper, while the shorter wavelength side is broader.[3,7] Thus, the actual UV curve deviates from Gaussian distribution. However, for the purpose of approximation by Gaussian distribution, the standard deviation $\Delta\sigma$ of the distribution was approximated by the average value of two actual $\Delta\sigma$ values on the longer and shorter wavelength sides, as indicated in Figure 10-2. In the present case, average $\Delta\sigma$ value of 2679.8 cm^{-1} was obtained from two $\Delta\sigma$ values 2272.9 and 3086.6 cm^{-1}. Twice the average $\Delta\sigma$ value is equal to the 1/e-width.

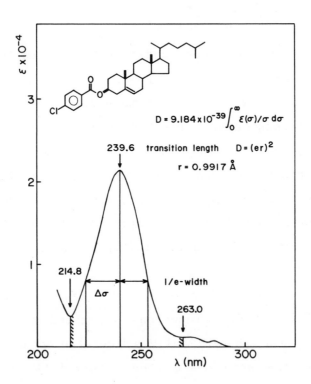

Figure 10-2.   *UV spectrum of cholesterol p-chlorobenzoate in ethanol.*

The empirical data[1] of transition length r and standard deviation $\Delta\sigma$ of pertinent <u>para</u>-substituted benzoates of cholesterol obtained from the observed UV spectra are tabulated in Table 10-1.

<u>Table 10-1.</u>   UV Data of <u>Para</u>-Substituted Benzoates of Cholesterol. [Reprinted from reference 1.]

Substituent	λ_{max}, nm	ε_{max}	r, Å	σ_{max}, cm^{-1}	$\Delta\sigma$, cm^{-1}
H	229.5	15 300	0.805	43 572.8	2769.0
Cl	240.0	21 400	0.992	41 666.6	2665.0
OMe	257.0	20 400	1.017	38 910.5	2886.1
NMe$_2$	311.4	31 600	1.294	32 113.0	2441.0
CN	240.0	24 600	1.068	41 666.6	2682.1
NO$_2$	260.5	15 100	1.034	38 387.7	3869.3

[a] UV spectra were measured in ethanol except for the nitrobenzoate (4% dioxane in ethanol). [b] In the methoxy, dimethylamino, and nitrobenzoates, the present transition length value contains that of a short axis polarized ^1L_b transition.

10-3.   Molecular Exciton Theory Applied to Binary System and Davydov Splitting[8,10]

The wave functions of a system composed of two or more chromophores were first derived for the ultraviolet absorption spectra of ionic and molecular crystals and this was developed into the molecular exciton theory by Davydov, in the late 1940's.[9]

In the case of a binary system composed of two identical chromophores i and j (Figure 10-3), exciton wave functions are defined as equations 10.8 and 10.9, in which each chromophore undergoes excitation 0→a, and no orbital overlap between two chromophores is assumed:

ground state: $\phi_{i0},\qquad \phi_{j0}$ (10.8)

excited state: $\phi_{ia},\qquad \phi_{ja}$ (10.9)

The Hamiltonian operator of the whole system is formulated as:

$$H = H_i + H_j + H_{ij}$$ (10.10)

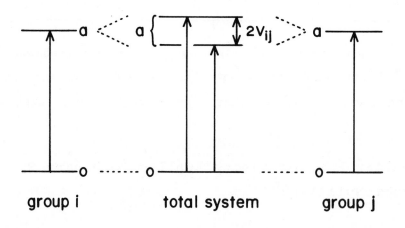

Figure 10-3. *Excited state splits into two energy levels by exciton interaction between two chromophores i and j. The energy gap* $2V_{ij}$ *is called the Davydov splitting.*

where H_i and H_j are the individual Hamiltonian of groups i and j, respectively, and H_{ij} is the interaction energy term between two groups i and j.

10-3-A. Wave Function and Energy of Ground State

The wave function of the ground state of the total system is expressed by the product of individual wave functions, as in equation 10.11.

ground state-wave function of total system:

$$\psi_0 = \phi_{i0}\phi_{j0} \tag{10.11}$$

The energy of the whole system in the ground state is calculable by operation on wave function ψ_0 with the Hamiltonian operator H and by integration over configurational space:

$$E = \int \psi_0 H \psi_0 d\tau$$

$$= \int \phi_{i0}\phi_{j0}[H_i + H_j + H_{ij}]\phi_{i0}\phi_{j0}d\tau$$

$$= \int \phi_{i0}H_i\phi_{i0}d\tau_i \int \phi_{j0}\phi_{j0}d\tau_j$$

$$+ \int \phi_{i0}\phi_{i0}d\tau_i \int \phi_{j0}H_j\phi_{j0}d\tau_j$$

$$+ \int \phi_{i0}\phi_{j0}H_{ij}\phi_{i0}\phi_{j0}d\tau \tag{10.12}$$

where the last term corresponds to the interaction between the two chromophores i and j in ground state, i.e., dipole–dipole interaction energy between the permanent dipoles of the chromophores.   In the case of nonpolar chromophores such as naphthalene and anthracene, the interaction term vanishes because of zero permanent dipole moment.   Similarly, in the case of polar chromophores, this term can be neglected because the term is relatively small and is effective only on the shift of absorption band.

Since wave functions ϕ_{i0} and ϕ_{j0} are normalized:

$$\int \phi_{i0}\phi_{i0}d\tau_i = \int \phi_{j0}\phi_{j0}d\tau_j = 1 \tag{10.13}$$

Therefore,

$$E = \int \phi_{i0}H_i\phi_{i0}d\tau_i$$

$$+ \int \phi_{j0}H_j\phi_{j0}d\tau_j$$

$$= 0 \tag{10.14}$$

Since the energy level of ground state of each chromophore is taken to be zero, the energy of the whole system in the ground state is zero.

10-3-B. Wave Function and Energy Level of Singly Excited States

The wave functions of singly excited states are also represented by the product of two individual wave functions. For example, the wave function of a total system composed of chromophore i in the excited state a and chromophore j in the ground state 0 is formulated as the product of two wave functions ϕ_{ia} and ϕ_{j0}.

$$\text{excited wave function:} \qquad \phi_{ia}\phi_{j0} \qquad\qquad (10.15)$$

In the case of the singly excited state with excited j chromophore, the wave function is defined as follows.

$$\text{excited wave function:} \qquad \phi_{i0}\phi_{ja} \qquad\qquad (10.16)$$

Since groups i and j are identical to each other, these two singly excited states are in a degenerate energy level. This degeneracy is resolved by the mutual interaction between chromophores i and j. Because of the assumption that orbital overlap between the two chromophores is negligible, the interaction can be approximated by a dipole-dipole interaction.

The wave function of the total system is represented by the linear combination of two wave functions of equations 10.15 and 10.16:

$$\psi_a = C_{iaj0}\phi_{ia}\phi_{j0} + C_{i0ja}\phi_{i0}\phi_{ja} \qquad\qquad (10.17)$$

The coefficients and energy levels are calculable by the variation method, just as in the case of the molecular orbital calculation of ethylene molecule. The second order secular equation is formulated as follows:

$$
\begin{vmatrix}
H_{iaj0;iaj0} - E & H_{iaj0;i0ja} - E\,S_{iaj0;i0ja} \\
\\
H_{iaj0;i0ja} - E\,S_{iaj0;i0ja} & H_{i0ja;i0ja} - E
\end{vmatrix} = 0
$$

$$(10.18)$$

where

$$H_{iaj0;iaj0} = \int \phi_{ia}\phi_{j0} H \phi_{ia}\phi_{j0} d\tau \qquad\qquad (10.19)$$

$$H_{i0ja;i0ja} = \int \phi_{i0}\phi_{ja} H \phi_{i0}\phi_{ja} d\tau \qquad\qquad (10.20)$$

$$H_{iaj0;i0ja} = \int \phi_{ia}\phi_{j0} H \phi_{i0}\phi_{ja} d\tau \qquad\qquad (10.21)$$

$$S_{iaj0;i0ja} = \int \phi_{ia}\phi_{j0}\phi_{i0}\phi_{ja} d\tau$$

$$\qquad\qquad = \int \phi_{ia}\phi_{i0} d\tau_i \int \phi_{j0}\phi_{ja} d\tau_j = 0 \qquad\qquad (10.22)$$

Since wave functions of ground and excited states of each chromophore — eg., ϕ_{i0} and ϕ_{ia} — are orthogonal to each other, the term $S_{iaj0;i0ja}$ in equation 10.22 is zero. Thus, the secular equation is simplified as

$$
\begin{vmatrix}
H_{iaj0;iaj0} - E & H_{iaj0;i0ja} \\
\\
H_{iaj0;i0ja} & H_{i0ja;i0ja} - E
\end{vmatrix} = 0 \qquad\qquad (10.23)
$$

The diagonal term $H_{iaj0;iaj0}$ is calculable as follows:

$$H_{iaj0;iaj0} = \int \phi_{ia}\phi_{j0}[H_i + H_j + H_{ij}]\phi_{ia}\phi_{j0}d\tau$$

$$= \int \phi_{ia}H_i\phi_{ia}d\tau_i \int \phi_{j0}\phi_{j0}d\tau_j$$

$$+ \int \phi_{ia}\phi_{ia}d\tau_i \int \phi_{j0}H_j\phi_{j0}d\tau_j$$

$$+ \int \phi_{ia}\phi_{j0}H_{ij}\phi_{ia}\phi_{j0}d\tau \qquad\qquad (10.24)$$

Because of the normalization of wave functions ϕ_{ia} and ϕ_{j0}, equation 10.24 becomes

$$H_{iaj0;iaj0} = \int \phi_{ia}H_i\phi_{ia}d\tau_i$$

$$+ \int \phi_{j0}H_j\phi_{j0}d\tau_j$$

$$+ \int \phi_{ia}\phi_{j0}H_{ij}\phi_{ia}\phi_{j0}d\tau \qquad\qquad (10.25)$$

where the first term is the energy of excited state a of chromophore i and is defined as E_a, and the second term is the energy of ground state of chromophore j and becomes zero as indicated by equation 10.14. The last term is the interaction energy between chromophore i in the excited state a and chromophore j in ground state 0, i.e., dipole-dipole interaction energy between two permanent dipoles of excited and ground states, and vanishes in the case of nonpolar chromophores. This term is negligible also in the case of polar

chromophores, as discussed in section 10-3-A.   Thus,   the diagonal   term is
simplified to

$$H_{iaj0;iaj0} = E_a \qquad\qquad\qquad (10.26)$$

Similarly, the other diagonal term $H_{i0ja;i0ja}$ is

$$H_{i0ja;i0ja} = E_a \qquad\qquad\qquad (10.27)$$

The off-diagonal term is formulated as follows:

$$H_{iaj0;i0ja} = \int \phi_{ia} \phi_{j0} [H_i + H_j + H_{ij}] \phi_{i0} \phi_{ja} d\tau$$

$$= \int \phi_{ia} H_i \phi_{i0} d\tau_i \int \phi_{j0} \phi_{ja} d\tau_j$$

$$+ \int \phi_{ia} \phi_{i0} d\tau_i \int \phi_{j0} H_j \phi_{ja} d\tau_j$$

$$+ \int \phi_{ia} \phi_{j0} H_{ij} \phi_{i0} \phi_{ja} d\tau \qquad\qquad (10.28)$$

The first and second terms vanish because of orthogonality of chromophoric wave functions, as exemplified by equation 10.22.   Thus, the off-diagonal
term is simplified:

$$H_{iaj0;i0ja} = \int \phi_{ia} \phi_{j0} H_{ij} \phi_{i0} \phi_{ja} d\tau \qquad\qquad (10.29)$$

This term corresponds to the dipole-dipole interaction energy between two
transition dipole moments of chromophores i and j.   The transition dipole
moments are due to the charge density polarization in the chromophore gener-
ated by the excitation $0 \rightarrow a$, and are proportional to the UV intensity.

Therefore, $H_{iaj0;i0ja}$ has a non-zero value, V_{ij}, and is approximated by a point dipole approximation method:

$$H_{iaj0;i0ja} = V_{ij} \tag{10.30}$$

$$V_{ij} = \mu_{i0a}\mu_{j0a}R_{ij}^{-3}\{\vec{e}_i \cdot \vec{e}_j - 3(\vec{e}_i \cdot \vec{e}_{ij})(\vec{e}_j \cdot \vec{e}_{ij})\} \tag{10.31}$$

where μ_{i0a} and μ_{j0a} are electric transition moments of groups i and j: R_{ij} is the distance between the two dipole moments; $\vec{e}_i$, $\vec{e}_j$, and $\vec{e}_{ij}$ are unit vectors of $\vec{\mu}_{i0a}$, $\vec{\mu}_{j0a}$, and $\vec{R}_{ij}$, respectively.

Then, the resultant secular equation is

$$\begin{vmatrix} E_a - E & V_{ij} \\ \\ V_{ij} & E_a - E \end{vmatrix} = 0 \tag{10.32}$$

$$(E_a - E)^2 - V_{ij}^2 = 0 \tag{10.33}$$

From this equation, the following eigen values and eigen functions are obtained:

α-state

$$E^{\alpha} = E_a - V_{ij} \qquad \psi_a^{\alpha} = (1/\sqrt{2})\{\phi_{ia}\phi_{j0} - \phi_{i0}\phi_{ja}\} \tag{10.34}$$

β-state

$$E^{\beta} = E_a + V_{ij} \qquad \psi_a{}^{\beta} = (1/\sqrt{2})\{\phi_{ia}\phi_{j0} + \phi_{i0}\phi_{ja}\} \qquad (10.35)$$

The present results represent the basic concepts of molecular exciton and the Davydov splitting. Namely, the dipole-dipole interaction between two transition dipoles splits the singly excited state of binary systems into two states α and β, as illustrated in Figure 10-3; this is the exciton coupling mechanism. The energy gap between α- and β-states expressed by $2V_{ij}$ is called the Davydov splitting after the name of the discoverer of the splitting.[8,9]

10-4. Electric Transition Moment of a Binary System

In order to calculate the rotational strength by the Rosenfeld equation (10.2), it is necessary to evaluate electric $\langle 0 | \vec{\mu} | a \rangle$ and magnetic $\langle a | \vec{M} | 0 \rangle$ transition moments by employing exciton wavefunctions ψ_0, $\psi_a{}^{\alpha}$, and $\psi_a{}^{\beta}$.

The electric dipole moment operator $\vec{\mu}$ of the total system is defined as

$$\vec{\mu} = \sum_i \vec{\mu}_i$$

$$\vec{\mu}_i = \sum_s e\vec{r}_{is} \qquad (10.36)$$

where $\vec{\mu}_i$ is the electric dipole moment operator of group i, and $\vec{r}_{is}$ is the distance vector of electron s in group i from the origin.

Electric transition moment is origin-independent as indicated below:

$$\vec{\mu}_{0a} = \int \psi_0 \sum_s e \vec{r}_s \psi_a d\tau$$

$$= \int \psi_0 \sum_s e (\vec{R} + \vec{r}_s') \psi_a d\tau \tag{10.37}$$

where $\vec{R}$ is a constant distance vector from origin to group i (Figure 10-4).

$$\vec{\mu}_{0a} = \sum_s e \vec{R} \int \psi_0 \psi_a d\tau + \int \psi_0 \sum_s e \vec{r}_s' \psi_a d\tau \tag{10.38}$$

The first term vanishes because of the orthogonality of functions ψ_0 and ψ_a:

$$\vec{\mu}_{0a} = \int \psi_0 \sum_s e \vec{r}_s' \psi_a d\tau \tag{10.39}$$

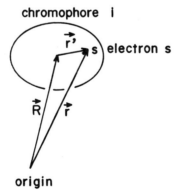

Figure 10-4. *Distance vector $\vec{r}$ is divided into two parts, $\vec{R}$ and $\vec{r}'$.*

Thus, the electric transition moment is origin-independent.

In the case of the permanent dipole moment of a neutral molecule, permanent dipole moment is expressed by

$$\vec{\mu} = -e\int \psi_0 \sum_s \vec{r}_s \psi_0 d\tau + e\sum_i z_i \vec{r}_i \qquad (10.40)$$

where Z_i and $\vec{r}_i$ are charge and position vector of nucleus i, respectively. It is easily seen that permanent dipole moment is also origin-independent.

The electric transition moment $\vec{\mu}_{0a}^{\alpha}$ of the excitation from ground state to the α excited state in a total system is formulated as:

α-state

$$<0|\vec{\mu}|a>^{\alpha} = \vec{\mu}_{0a}^{\alpha} = \int \psi_0 \vec{\mu} \psi_a^{\alpha} d\tau$$

$$= (1/\sqrt{2}) \int \phi_{i0}\phi_{j0}\vec{\mu}(\phi_{ia}\phi_{j0} - \phi_{i0}\phi_{ja}) d\tau \qquad (10.41)$$

The electric dipole moment operator $\vec{\mu}$ can be divided into two groups i and j:

$$\vec{\mu} = \vec{\mu}_i + \vec{\mu}_j \qquad (10.42)$$

Therefore,

$$\vec{\mu}_{0a}{}^{\alpha} = (1/\sqrt{2}) \int \phi_{i0}\phi_{j0} [\vec{\mu}_i + \vec{\mu}_j] (\phi_{ia}\phi_{j0} - \phi_{i0}\phi_{ja}) d\tau$$

$$= (1/\sqrt{2}) \{ \int \phi_{i0}\vec{\mu}_i \phi_{ia} d\tau_i \int \phi_{j0}\phi_{j0} d\tau_j$$

$$+ \int \phi_{i0}\phi_{ia} d\tau_i \int \phi_{j0}\vec{\mu}_j \phi_{j0} d\tau_j$$

$$- \int \phi_{i0}\vec{\mu}_i \phi_{i0} d\tau_i \int \phi_{j0}\phi_{ja} d\tau_j$$

$$- \int \phi_{i0}\phi_{i0} d\tau_i \int \phi_{j0}\vec{\mu}_j \phi_{ja} d\tau_j \} \qquad (10.43)$$

where the second and third terms vanish.

$$\vec{\mu}_{0a}{}^{\alpha} = (1/\sqrt{2}) \{ \int \phi_{i0}\vec{\mu}_i \phi_{ia} d\tau_i - \int \phi_{j0}\vec{\mu}_j \phi_{ja} d\tau_j \}$$

$$= (1/\sqrt{2}) (\vec{\mu}_{i0a} - \vec{\mu}_{j0a}) \qquad (10.44)$$

where

$$\vec{\mu}_{i0a} = \int \phi_{i0}\vec{\mu}_i \phi_{ia} d\tau_i \qquad (10.45)$$

$$\vec{\mu}_{j0a} = \int \phi_{j0}\vec{\mu}_j \phi_{ja} d\tau_j \qquad (10.46)$$

In a similar way, the electric transition moment $\vec{\mu}_{0a}{}^{\beta}$ of the excitation $0 \rightarrow \beta$ is derived as follows:

$$\vec{\mu}_{0a}{}^{\beta} = (1/\sqrt{2}) (\vec{\mu}_{i0a} + \vec{\mu}_{j0a}) \qquad (10.47)$$

The present results reveal the important properties of two transition dipole moments $\vec{\mu}_{0a}^{\,\alpha}$ and $\vec{\mu}_{0a}^{\,\beta}$. Namely, as easily understood from Figure 10-5 and equations 10.44 and 10.47, dipole moments are perpendicular to each other, and if the angle between two dipole moments $\vec{\mu}_{i0a}$ and $\vec{\mu}_{j0a}$ is less than 90°, the magnitude of $\vec{\mu}_{0a}^{\,\beta}$ is larger than that of $\vec{\mu}_{0a}^{\,\alpha}$. Therefore, in this case, the UV intensity of the β-state is stronger than that of the α-state. In the region larger than 90°, $\vec{\mu}_{0a}^{\,\alpha}$ is larger than $\vec{\mu}_{0a}^{\,\beta}$, and hence the α-state exhibits a more intense UV absorption band than the β-state.

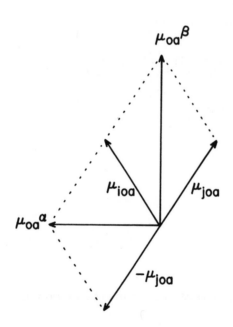

Figure 10-5. *Electric transition dipole moments of α- and β-states are perpendicular to each other.*

10-5.   Magnetic Transition Moment of a Binary System

The magnetic moment operator $\vec{M}$ of a whole system is formulated as

$$\vec{M} = \sum_i \vec{M}_i \tag{10.48}$$

$$\vec{M}_i = (e/2mc) \sum_s (\vec{r}_{is} \times \vec{p}_{is}) \tag{10.49}$$

where m is mass of electron, c is velocity of light, $\vec{r}_{is}$ is distance vector of electron s in group i from the origin, $\vec{p}_{is}$ is linear momentum of electron s in group i, and × stands for vector product of two vectors.

Since

$$\vec{p}_{is} = (h/2\pi i) (\vec{i}\frac{\partial}{\partial x} + \vec{j}\frac{\partial}{\partial y} + \vec{k}\frac{\partial}{\partial z}) \tag{10.50}$$

the magnetic moment operator is also expressed in rectangular coordinate as[11]

$$\vec{M}_{is} = (\frac{eh}{4\pi mci}) \{\vec{i}\,(y\frac{\partial}{\partial z} - z\frac{\partial}{\partial y}) + \vec{j}\,(z\frac{\partial}{\partial x} - x\frac{\partial}{\partial z}) + \vec{k}\,(x\frac{\partial}{\partial y} - y\frac{\partial}{\partial x})\} \tag{10.51}$$

where h is Planck's constant.

In analogy with the case of electric transition moment, the distance vector is divided into two parts as illustrated in Figure 10-6.

$$\vec{r}_{is} = \vec{R}_i + \vec{r}_{is}' \tag{10.52}$$

where $\vec{R}_i$ is distance vector of group i from the origin.   Therefore, the magnetic moment operator is also divided into two terms:

$$\vec{M}_i = (e/2mc)\sum_s \vec{R}_i \times \vec{p}_{is} + (e/2mc)\sum_s \vec{r}_{is}' \times \vec{p}_{is}$$

$$= (e/2mc)\vec{R}_i \times \sum_s \vec{p}_{is} + (e/2mc)\sum_s \vec{r}_{is}' \times \vec{p}_{is}$$

$$= (e/2mc)\vec{R}_i \times \vec{p}_i + \vec{m}_i \tag{10.53}$$

where $\vec{p}_i$ and $\vec{m}_i$ are linear momentum and internal magnetic moment operators of group i, respectively, and are formulated as follows:

$$\vec{p}_i = \sum_s \vec{p}_{is} \tag{10.54}$$

$$\vec{m}_i = \sum_s \vec{r}_{is}' \times \vec{p}_{is} \tag{10.55}$$

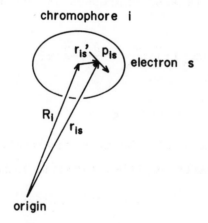

Figure 10-6. *Magnetic transition dipole moment vector is divided into two parts in analogy with the case of electric transition moment.*

Thus, magnetic moment operator $\vec{M}$ of the total system is:

$$\vec{M} = (e/2mc)\sum_i \vec{R}_i \times \vec{p}_i + \sum_i \vec{m}_i \qquad (10.56)$$

In binary systems composed of two chromophores i and j, the operator is divided into two parts:

$$\vec{M} = (e/2mc)\vec{R}_i \times \vec{p}_i + \vec{m}_i + (e/2mc)\vec{R}_j \times \vec{p}_j + \vec{m}_j \qquad (10.57)$$

The magnetic transition moment $\langle a \mid \vec{M} \mid 0 \rangle$ of excitation to α-state is calculated using the magnetic moment operator in equation 10.57:

$$\langle a|\vec{M}|0\rangle^\alpha = \int \psi_a^\alpha \vec{M} \psi_0 d\tau$$

$$= \int (1/\sqrt{2})(\phi_{ia}\phi_{j0} - \phi_{i0}\phi_{ja})\vec{M}\phi_{i0}\phi_{j0}d\tau$$

$$= (1/\sqrt{2})\int(\phi_{ia}\phi_{j0} - \phi_{i0}\phi_{ja})[(e/2mc)\vec{R}_i \times \vec{p}_i + \vec{m}_i$$

$$+ (e/2mc)\vec{R}_j \times \vec{p}_j + \vec{m}_j]\phi_{i0}\phi_{j0}d\tau$$

$$= (1/\sqrt{2})\{(e/2mc)\int\phi_{ia}\phi_{j0}(\vec{R}_i \times \vec{p}_i)\phi_{i0}\phi_{j0}d\tau$$

$$- (e/2mc)\int\phi_{i0}\phi_{ja}(\vec{R}_i \times \vec{p}_i)\phi_{i0}\phi_{j0}d\tau$$

$$+ \int\phi_{ia}\phi_{j0}\vec{m}_i\phi_{i0}\phi_{j0}d\tau$$

$$-\int \phi_{i0}\phi_{ja}\vec{m}_i\phi_{i0}\phi_{j0}d\tau$$

$$+(e/2mc)\int \phi_{ia}\phi_{j0}(\vec{R}_j\times\vec{p}_j)\phi_{i0}\phi_{j0}d\tau$$

$$-(e/2mc)\int \phi_{i0}\phi_{ja}(\vec{R}_j\times\vec{p}_j)\phi_{i0}\phi_{j0}d\tau$$

$$+\int \phi_{ia}\phi_{j0}\vec{m}_j\phi_{i0}\phi_{j0}d\tau$$

$$-\int \phi_{i0}\phi_{ja}\vec{m}_j\phi_{i0}\phi_{j0}d\tau\}\tag{10.58}$$

where second, fourth, fifth, and seventh terms vanish because of orthogonality of wave functions.

Therefore,

$$<a|\vec{M}|0>^\alpha = (1/\sqrt{2})\{(e/2mc)\vec{R}_i\times\int \phi_{ia}\vec{p}_i\phi_{i0}d\tau_i$$

$$+\int \phi_{ia}\vec{m}_i\phi_{i0}d\tau_i$$

$$-(e/2mc)\vec{R}_j\times\int \phi_{ja}\vec{p}_j\phi_{j0}d\tau_j$$

$$-\int \phi_{ja}\vec{m}_j\phi_{j0}d\tau_j\}$$

$$= (1/\sqrt{2})\{(e/2mc)\vec{R}_i\times\vec{p}_{ia0}+\vec{m}_{ia0}$$

$$-(e/2mc)\vec{R}_j\times\vec{p}_{ja0}-\vec{m}_{ja0}\}\tag{10.59}$$

where $\vec{p}_{ia0}$ and $\vec{m}_{ia0}$ are linear momentum and internal magnetic moment of group i, respectively.

$$\vec{P}_{ia0} = \int \phi_{ia} \vec{P}_i \phi_{i0} d\tau_i \tag{10.60}$$

$$\vec{m}_{ia0} = \int \phi_{ia} \vec{m}_i \phi_{i0} d\tau_i \tag{10.61}$$

$\vec{P}_{ja0}$ and $\vec{m}_{ja0}$ are similarly defined.

For the linear momentum of a group, the following useful equation was derived.[12]

$$\vec{P}_{0a} = -(2\pi imc/e) \sigma_0 \vec{\mu}_{0a} \tag{10.62}$$

where σ_0 and $\vec{\mu}_{0a}$ are excitation wave number and electric transition moment of transition $0 \rightarrow a$, respectively, and are formulated as follows:

$$\vec{P}_{0a} = \int \phi_0 \vec{p} \phi_a d\tau \tag{10.63}$$

$$\sigma_0 = \nu_0/c = (E_a - E_0)/(hc) \tag{10.64}$$

$$\vec{\mu}_{0a} = \int \phi_0 \vec{\mu} \phi_a d\tau \tag{10.65}$$

Since linear momentum operator $\vec{p}$ is Hermitian and imaginary,

$$\vec{P}_{a0} = -\vec{P}_{0a} \tag{10.66}$$

If an operator α satisfies the following equation for any two functions ψ and ϕ, the operator α is said to be Hermitian:[13]

$$\int \phi^*(\alpha\psi)d\tau = \int \psi(\alpha^*\phi^*)d\tau \qquad (10.67)$$

where * denotes complex conjugate.

From equations 10.63 and 10.67,

$$\vec{P}_{0a} = \int \phi_0^*(\vec{p}\phi_a)d\tau = \int \phi_a(\vec{p}^*\phi_0^*)d\tau \qquad (10.68)$$

Since linear momentum operator $\vec{p}$ is imaginary as expressed:

$$\vec{p} = (h/2\pi i)(\vec{i}\frac{\partial}{\partial x} + \vec{j}\frac{\partial}{\partial y} + \vec{k}\frac{\partial}{\partial z}) \qquad (10.69)$$

$$\vec{p}^* = -(h/2\pi i)(\vec{i}\frac{\partial}{\partial x} + \vec{j}\frac{\partial}{\partial y} + \vec{k}\frac{\partial}{\partial z}) \qquad (10.70)$$

Since each wave function is chosen to be real,

$$\vec{P}_{0a} = \int \phi_a \vec{p}^* \phi_0 d\tau$$

$$= \int \phi_a(-\vec{p})\phi_0 d\tau$$

$$= -\int \phi_a \vec{p}\phi_0 d\tau$$

$$= -\vec{P}_{a0} \qquad (10.71)$$

Thus, equation 10.66 is derived.

From equations 10.62 and 10.66

$$\vec{P}_{ia0} = (2\pi imc/e)\sigma_0\vec{\mu}_{i0a} \tag{10.72}$$

Thus, the transition linear momentum $\vec{P}_{ia0}$ of chromophore i is correlated with the electric transition moment $\vec{\mu}_{i0a}$. According to equations 10.59 and 10.72,

$$\langle a|\vec{M}|0\rangle^\alpha = (1/\sqrt{2})\{(e/2mc)\vec{R}_i\times(2\pi imc/e)\sigma_0\vec{\mu}_{i0a}+\vec{m}_{ia0}$$

$$-(e/2mc)\vec{R}_j\times(2\pi imc/e)\sigma_0\vec{\mu}_{j0a}-\vec{m}_{ja0}\}$$

$$= (1/\sqrt{2})\{i\pi\sigma_0\vec{R}_i\times\vec{\mu}_{i0a}-i\pi\sigma_0\vec{R}_j\times\vec{\mu}_{j0a}+\vec{m}_{ia0}-\vec{m}_{ja0}\} \tag{10.73}$$

In a similar way, the magnetic transition moment of excitation to β-state is calculable:

$$\langle a|\vec{M}|0\rangle^\beta = (1/\sqrt{2})\{i\pi\sigma_0\vec{R}_i\times\vec{\mu}_{i0a}+i\pi\sigma_0\vec{R}_j\times\vec{\mu}_{j0a}+\vec{m}_{ia0}+\vec{m}_{ja0}\} \tag{10.74}$$

As indicated in equations 10.73 and 10.74, magnetic transition moments expressed in these equations depend on the origin.

10-6. Rotational Strength of a Binary System

The rotational strength of a binary system now can be calculated, as follows, using equations 10.2, 10.44, 10.47, 10.73, and 10.74. In the case of α-state:

$$R^{\alpha} = \text{Im}\{<0|\vec{\mu}|a>^{\alpha} \cdot <a|\vec{M}|0>^{\alpha}\}$$

$$= \text{Im}\{(1/\sqrt{2})(\vec{\mu}_{i0a}-\vec{\mu}_{j0a}) \cdot (1/\sqrt{2})(\vec{m}_{ia0}-\vec{m}_{ja0}$$

$$+i\pi\sigma_0\vec{R}_i\times\vec{\mu}_{i0a}-i\pi\sigma_0\vec{R}_j\times\vec{\mu}_{j0a})\}$$

$$= (1/2)\text{Im}\{(\vec{\mu}_{i0a}-\vec{\mu}_{j0a}) \cdot (\vec{m}_{ia0}-\vec{m}_{ja0})\}$$

$$+(1/2)\pi\sigma_0(\vec{\mu}_{i0a}-\vec{\mu}_{j0a}) \cdot (\vec{R}_i\times\vec{\mu}_{i0a}-\vec{R}_j\times\vec{\mu}_{j0a})$$

$$= (1/2)\text{Im}\{(\vec{\mu}_{i0a}-\vec{\mu}_{j0a}) \cdot (\vec{m}_{ia0}-\vec{m}_{ja0})\}$$

$$+(1/2)\pi\sigma_0\{\vec{\mu}_{i0a} \cdot (\vec{R}_i\times\vec{\mu}_{i0a})-\vec{\mu}_{i0a} \cdot (\vec{R}_j\times\vec{\mu}_{j0a})$$

$$-\vec{\mu}_{j0a} \cdot (\vec{R}_i\times\vec{\mu}_{i0a})+\vec{\mu}_{j0a} \cdot (\vec{R}_j\times\vec{\mu}_{j0a})\} \tag{10.75}$$

the second term is simplified by using the following general equations for vectors:

$$\vec{A} \cdot (\vec{B}\times\vec{C}) = \vec{B} \cdot (\vec{C}\times\vec{A}) \tag{10.76}$$

$$\vec{A} \times \vec{A} = 0 \tag{10.77}$$

$$\vec{A} \cdot (\vec{B} \times \vec{A}) = \vec{B} \cdot (\vec{A} \times \vec{A}) = 0 \qquad (10.78)$$

Hence,

$$R^{\alpha} = (1/2) \operatorname{Im}\{ (\vec{\mu}_{i0a} - \vec{\mu}_{j0a}) \cdot (\vec{m}_{ia0} - \vec{m}_{ja0}) \}$$

$$+ (1/2) \pi \sigma_0 \{ \vec{R}_j \cdot (\vec{\mu}_{i0a} \times \vec{\mu}_{j0a}) - \vec{R}_i \cdot (\vec{\mu}_{i0a} \times \vec{\mu}_{j0a}) \}$$

$$= (1/2) \operatorname{Im}\{ (\vec{\mu}_{i0a} - \vec{\mu}_{j0a}) \cdot (\vec{m}_{ia0} - \vec{m}_{ja0}) \}$$

$$+ (1/2) \pi \sigma_0 \vec{R}_{ij} \cdot (\vec{\mu}_{i0a} \times \vec{\mu}_{j0a}) \qquad (10.79)$$

where $\vec{R}_{ij}$ is the interchromophoric distance vector from group i to group j, and expressed as:

$$\vec{R}_{ij} = \vec{R}_j - \vec{R}_i \qquad (10.80)$$

In a similar way, the rotational strength of β-state is formulated as

$$R^{\beta} = \operatorname{Im}\{ (1/\sqrt{2}) (\vec{\mu}_{i0a} + \vec{\mu}_{j0a}) \cdot (1/\sqrt{2}) (\vec{m}_{ia0} + \vec{m}_{ja0}$$

$$+ i \pi \sigma_0 \vec{R}_i \times \vec{\mu}_{i0a} + i \pi \sigma_0 \vec{R}_j \times \vec{\mu}_{j0a}) \}$$

$$= (1/2) \operatorname{Im}\{ (\vec{\mu}_{i0a} + \vec{\mu}_{j0a}) \cdot (\vec{m}_{ia0} + \vec{m}_{ja0}) \}$$

$$+ (1/2) \pi \sigma_0 \{ \vec{\mu}_{i0a} \cdot \vec{R}_j \times \vec{\mu}_{j0a} + \vec{\mu}_{j0a} \cdot \vec{R}_i \times \vec{\mu}_{i0a} \}$$

$$= (1/2) \operatorname{Im}\{ (\vec{\mu}_{i0a} + \vec{\mu}_{j0a}) \cdot (\vec{m}_{ia0} + \vec{m}_{ja0}) \}$$

$$-(1/2)\pi\sigma_0\vec{R}_{ij} \cdot (\vec{\mu}_{i0a} \times \vec{\mu}_{j0a}) \qquad\qquad (10.81)$$

In equations 10.79 and 10.81, each first term corresponds to the rotational strength due to internal magnetic transition moments, while the second term denotes the rotational strength based on dipole-dipole coupling between two groups i and j; i.e., chiral exciton coupling. In the case of $\pi \rightarrow \pi^*$ transition of common molecules, internal magnetic transition moments $\vec{m}_{ia0}$ and $\vec{m}_{ja0}$ are zero or relatively small. Therefore, the first term is negligible in comparison with the second term. Thus, rotational strength is approximated as

$$R^{\alpha,\beta} = \pm(1/2)\pi\sigma_0\vec{R}_{ij} \cdot (\vec{\mu}_{i0a} \times \vec{\mu}_{j0a}) \qquad\qquad (10.82)$$

where upper and lower signs correspond to α- and β-states, respectively.

Equation 10.82 indicates the following significant and characteristic properties of exciton split Cotton effects:

1. The Cotton effects of α- and β-states have identical rotational strengths of opposite signs. Namely, two split Cotton effects are conservative, and satisfy the sum rule:

$$\sum_A R^A = 0 \qquad\qquad (10.83)$$

2. The rotational strength is proportional to the triple product of interchromophoric distance and electric transition moments of groups i and j. Therefore, exciton coupling between intense

π→π* transitions generates strong Cotton effects. Thus, in order to observe clear exciton split Cotton effects for determining the absolute stereochemistry, it is advisable to use chromophores undergoing intense π→π* transition.

3. Since rotational strength is a physically observable quantity as indicated in equation 10.1, rotational strength should be origin-independent. However, because of the origin-dependence of magnetic moment operator, it is, in general, rather difficult to obtain origin-independent results in theoretical calculations. In the case of exciton coupling, as indicated in equation 10.82, the formula of rotational strength includes only the interchromophoric distance vector which is origin-independent. Thus, equation 10.82 satisfies the origin-independence of rotational strength.

10 - 7. Theoretical Derivation of Exciton Chirality Method[14]

The theoretical calculation results discussed in the previous sections are summarized in Table 10-2.

From these equations, the following generality can be derived: when the equations are applied to a C_2-symmetrical system, equations 10.87 and 10.91 indicate that the α-state corresponds to B-symmetry, and the β-state to A-symmetry. If the transitions of the system make a right-handed helicity, i.e., $\vec{R}_{ij} \cdot (\vec{\mu}_{i0a} \times \vec{\mu}_{j0a}) > 0$, the rotational

<u>**Table 10–2.**</u> **Exciton Wave Function, Excitation Energy, and Dipole and Rotational Strengths of Binary Systems.**

$$V_{ij} = \mu_{i0a}\mu_{j0a}R_{ij}^{-3}\{\vec{e}_i \cdot \vec{e}_j - 3(\vec{e}_i \cdot \vec{e}_{ij})(\vec{e}_j \cdot \vec{e}_{ij})\}$$

$$= R_{ij}^{-3}\{\vec{\mu}_{i0a} \cdot \vec{\mu}_{j0a} - 3R_{ij}^{-2}(\vec{\mu}_{i0a} \cdot \vec{R}_{ij})(\vec{\mu}_{j0a} \cdot \vec{R}_{ij})\} \qquad (10.84)$$

α-sate

$$\psi_a^{\alpha} = (1/\sqrt{2})(\phi_{ia}\phi_{j0} - \phi_{i0}\phi_{ja}) \qquad (10.85)$$

$$E^{\alpha} = E_a - V_{ij} \qquad (10.86)$$

$$D^{\alpha} = (1/2)(\vec{\mu}_{i0a} - \vec{\mu}_{j0a})^2 \qquad (10.87)$$

$$R^{\alpha} = +(1/2)\pi\sigma_0\vec{R}_{ij} \cdot (\vec{\mu}_{i0a} \times \vec{\mu}_{j0a}) \qquad (10.88)$$

β-state

$$\psi_a^{\beta} = (1/\sqrt{2})(\phi_{ia}\phi_{j0} + \phi_{i0}\phi_{ja}) \qquad (10.89)$$

$$E^{\beta} = E_a + V_{ij} \qquad (10.90)$$

$$D^{\beta} = (1/2)(\vec{\mu}_{i0a} + \vec{\mu}_{j0a})^2 \qquad (10.91)$$

$$R^{\beta} = -(1/2)\pi\sigma_0\vec{R}_{ij} \cdot (\vec{\mu}_{i0a} \times \vec{\mu}_{j0a}) \qquad (10.92)$$

strength of B-symmetrical transition is positive, as derived from equation 10.88. Namely, the B-symmetrical transition generates a positive Cotton effect. On the other hand, the A-symmetrical one gives a negative CD. If the system is composed of left-handed helicity of transition moments, i.e., $\vec{R}_{ij} \cdot (\vec{\mu}_{i0a} \times \vec{\mu}_{j0a}) < 0$, the sign is reversed. This is a modification of the so-called "C_2-symmetrical rule."[15]

Now, in order to prove the conclusion of the exciton chirality method, these equations are applied to model systems of vicinal glycol dibenzoate.

10-7-A. Dibenzoate Model System with Positive Exciton Chirality

The first model system is a vicinal glycol dibenzoate with right-handed or clockwise screwness, as illustrated in Figure 10-7. In this system, one can arbitrarily choose one of four combination modes of two transition dipole moments: i.e., two in-phase combinations and two out-of-phase combinations.

The case with in-phase combination.

Now, let us consider the first case with in-phase combination as shown in Figure 10-7. In this model, dipole-dipole interaction energy V_{ij} expressed by equation 10.84 is calculated to be positive.

As shown in detail in section 11-2, the first term of (10.84) is positive because the angle between the two dipole moments is less than 90°, while the second term is always positive irrespective of values of the angles involved. Thus, it is clear that the value of dipole-dipole interaction energy V_{ij} is positive.

Since the V_{ij} value is positive, the α-state corresponds to longer wavelength transition, i.e, a first Cotton effect, and the β-state corresponds to shorter wavelength transition, i.e., a second Cotton effect.

Now, the vector product between $\vec{\mu}_i$ and $\vec{\mu}_j$ generates a vector $(\vec{\mu}_{i0a} \times \vec{\mu}_{j0a})$ which is almost parallel to the interchromophoric distance vector $\vec{R}_{ij}$. Therefore, it is obvious that the triple product $\vec{R}_{ij} \cdot (\vec{\mu}_{i0a} \times \vec{\mu}_{j0a})$ takes a positive value. The equations in Table 10-2 indicate that the rotational strength of the α-state at longer wavelength is positive while that of the β-state   at

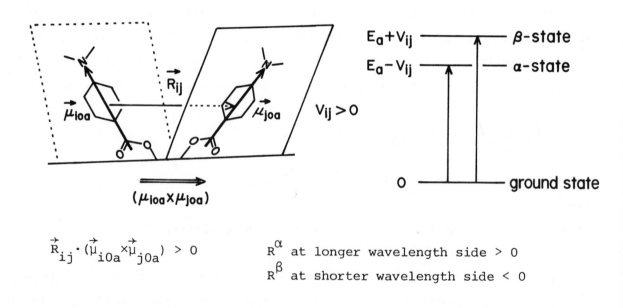

$$\vec{R}_{ij} \cdot (\vec{\mu}_{i0a} \times \vec{\mu}_{j0a}) > 0 \qquad\qquad R^{\alpha} \text{ at longer wavelength side} > 0$$
$$R^{\beta} \text{ at shorter wavelength side} < 0$$

Figure 10-7. *Positive exciton chirality model with in-phase combination.*

shorter wavelength is negative.   Thus, in line with the observed CD spectra, the sign of the first Cotton effect agrees with that of exciton chirality.

The case with out-of-phase combination.

Next, let us consider the second case with an out-of-phase combination, illustrated in Figure 10-8, in which the direction of the electric transition moment of group j is reversed.   In this case, dipole-dipole interaction energy V_{ij} is obviously negative because substitution of $-\vec{\mu}_{j0a}$ for $\vec{\mu}_{j0a}$ in equation 10.84 gives $-V_{ij}$.   Therefore, the β-state corresponds to the first Cotton effect and α-state to the second Cotton effect.   The rotational strength of

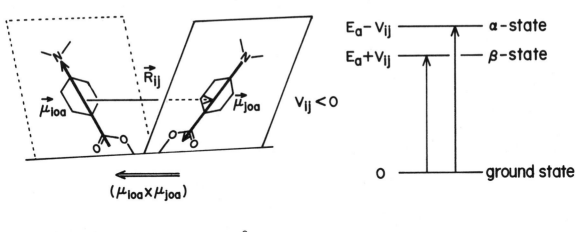

$$\vec{R}_{ij} \cdot (\vec{\mu}_{i0a} \times \vec{\mu}_{j0a}) < 0 \qquad \begin{array}{l} R^{\beta} \text{ at longer wavelength side} > 0 \\ R^{\alpha} \text{ at shorter wavelength side} < 0 \end{array}$$

Figure 10-8.   Positive exciton chirality model with out-of-phase combination.

β-state is positive, because vector product $\vec{\mu}_{i0a} \times \vec{\mu}_{j0a}$ is almost antiparallel to $\vec{R}_{ij}$; the triple product $\vec{R}_{ij} \cdot (\vec{\mu}_{i0a} \times \vec{\mu}_{j0a})$ is thus negative. On the other hand, the R value of α-state is necessarily negative. The calculated value of the first Cotton effect is hence positive while that of the second one is negative. This result is in accord with the positive sign of the exciton chirality.

Another case with out-of-phase combination.

The third case is another out-of-phase combination in which the direction of $\vec{\mu}_{i0a}$ is reversed. In analogy with the second case, dipole-dipole interaction energy V_{ij} is negative. Therefore, β-state corresponds to the first Cotton effect. Since vectors $(\vec{\mu}_{i0a} \times \vec{\mu}_{j0a})$ and $\vec{R}_{ij}$ are antiparallel to each other, triple product $\vec{R}_{ij} \cdot (\vec{\mu}_{i0a} \times \vec{\mu}_{j0a})$ is negative. The rotational strength of the β-state is positive while that of the α-state is negative. Thus, the positive sign of the first Cotton effect agrees with the positive exciton chirality.

Another case with in-phase combination.

The fourth case is another in-phase combination in which both transition moments are reversed in direction. When dipole moment vectors $\vec{\mu}_{i0a}$ and $\vec{\mu}_{j0a}$ in equation 10.84 are replaced by $-\vec{\mu}_{i0a}$ and $-\vec{\mu}_{j0a}$, respectively, the sign of V_{ij} remains unchanged. The same is true in the case of equations 10.88 and 10.92. Therefore, the situation is in complete accord with the first case. Thus, the sign of the first Cotton effect agrees with the sign of exciton chirality.

The above calculation results indicate the following significant nature of the exciton chirality method; namely, the sign of the first Cotton effect depends solely on the exciton chirality, irrespective of the initial choice of the phase of two vectors, i.e., the combination mode of two dipole moments.

Thus, the present theoretical treatments satisfy the requirements that observable physical value should be independent of initial arbitrary choice of phase.

In the course of electronic excitation, electrons oscillate to generate electric polarization which corresponds to the electric transition dipole moment; i.e., the electric transition moment has the nature of an oscillation. This is understandable as follows: if ψ^{ex} is a correct wave function of an excited state, $-\psi^{ex}$ is also a correct wave function. Therefore, both electric transition moments,

$$\int \psi_0 \vec{\mu} \psi^{ex} d\tau = \vec{\mu} \tag{10.93}$$

$$\int \psi_0 \vec{\mu} (-\psi^{ex}) d\tau = -\vec{\mu} \tag{10.94}$$

$\vec{\mu}$ and $-\vec{\mu}$ are correct, and have physical meanings.

Some organic chemists tend to confuse electric transition moment and electric permanent dipole moment. For example, in the case of p-dimethylaminobenzoate, the ground state has a permanent dipole moment which is directed from the amino group to the ester group. In the course of an $\pi \rightarrow \pi^*$ intramolecular charge transfer transition, electrons migrate from the amino to the ester group via the benzene ring. Therefore, the permanent dipole of the excited state is much

more polarized, and is directed as shown in Figure 10-9.
As shown in the following equations,

$$\int \psi^{ex}\vec{\mu}\psi^{ex}d\tau + e\sum_{i} z_i\vec{r}_i = \vec{\mu} \tag{10.95}$$

$$\int (-\psi^{ex})\vec{\mu}(-\psi^{ex})d\tau + e\sum_{i} z_i\vec{r}_i = \vec{\mu} \tag{10.96}$$

reversal in the sign of the wave function does not affect the sign of
the permanent dipole moment $\vec{\mu}$.   Thus,   the   permanent   dipole   is
expressed by a single arrow, the direction of which has a physical
meaning and indicates electric polarization from the electron donating
group to the electron withdrawing group.

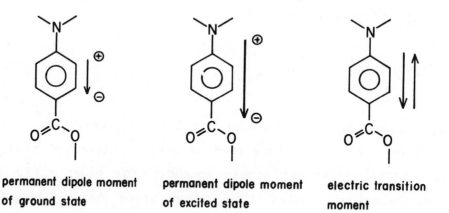

permanent dipole moment   permanent dipole moment   electric transition
of ground state   of excited state   moment

Figure 10-9. *Permanent dipole moments are expressed by a unidirectional*
arrow, while the electric transition moment is expressed by arrows pointing in
both directions.

On the other hand, the nature of electric transition moment is represented by two arrows indicating oscillation, as shown in Figure 10-9. Rotational strength of exciton Cotton effects depends only on electric transition moments and is independent of the permanent dipole moments. This is also verified by the model system of (6<u>R</u>,15<u>R</u>)-(+)-6,15-dihydro-6,15-ethanonaphtho[2,3-<u>c</u>]pentaphene (compound <u>4</u> in Chapter 1) having two nonpolar chromophores of anthracene.

10-7-B.   Dibenzoate Model System with Negative Exciton Chirality

Next, let us consider the vicinal glycol dibenzoate with negative exciton chirality. Theoretical treatments employed in the case of dibenzoate with positive exciton chirality hold for the present case in a similar way. The dipole-dipole interaction energy terms remain unchanged, while the triple product term $\vec{R}_{ij} \cdot (\vec{\mu}_{i0a} \times \vec{\mu}_{j0a})$ changes the sign because of the mirror image relationship. Therefore it is readily understandable that the first Cotton effect is negative and is consistent with the negative exciton chirality. These arguments prove the exciton chirality method in a qualitative sense.

10-7-C.   Model System of (6R,15R)-(+)-6,15-Dihydro-6,15-ethanonaphtho [2,3 -c]pentaphene with Positive Exciton Chirality[4]

The present compound (compound <u>4</u> in Chapter 1) is a C_2 symmetrical cage molecule with rigid skeleton. It is therefore an ideal case for a quantitative explanation of the exciton chirality method.   This compound exhibits typical and strong split Cotton effects in the ^1B_b transition region, the transition dipole moment of which runs along the long axis of the anthracene chromophore.   The phase of dipole moments is arbitrarily chosen as indicated in Figure 11-14, i.e., in-phase combination.   Since the angle between the two dipole moments is about 165°, the first term of equation 10.84 is negative.

$$\vec{e}_i \cdot \vec{e}_j = \cos 165° = -0.966 \tag{10.97}$$

On the other hand, since the angles between the interchromophoric distance vector $\vec{R}_{ij}$ and two dipole moments $\vec{\mu}_{i0a}$ and $\vec{\mu}_{j0a}$ are about 35° and 145°, respectively, the second term has the following positive value:

$$-3(\vec{e}_i \cdot \vec{e}_{ij})(\vec{e}_j \cdot \vec{e}_{ij}) = -3(\cos 35°)(\cos 145°)$$

$$= +2.013 \tag{10.98}$$

Thus, interaction energy V_{ij} is clearly positive.   The present positive value is also verified by quantitative calculations employing the point monopole approximation method.

Since V_{ij} is positive, the α-state is more stable and corresponds to the first Cotton effect, while the β-state of higher energy level corresponds to the second Cotton effect.   The present compound has C_2-symmetrical character.   Thus, as indicated in equation 10.85, the wave function of α-state   is anti-symmetrical against rotation around the C_2-axis, and the α-state   belongs to B-symmetry.   The wave function of β-state is symmetrical and therefore it belongs to A-symmetry (Figure 11-14).

Next, the triple product $\vec{R}_{ij} \cdot (\vec{\mu}_{i0a} \times \vec{\mu}_{j0a})$ governing the sign of rotational strength is positive in the present case because the two dipole moments constitute a clock-wise or right-handed screwness. Namely, a vector $(\vec{\mu}_{i0a} \times \vec{\mu}_{j0a})$ which is derived from the vector product of the two moments is almost parallel to the interchromophoric distance vector $\vec{R}_{ij}$. Thus, it is obvious that the rotational strength R^{α} of α-state is positive, while that of β-state is necessarily negative (Figure 11-14). Accordingly, the calculation results are in accordance with the observed CD data.

10-8.   CD of N-mer and Quantitative Definition of Exciton Chirality[3]

As seen in the previous sections, exciton chirality governs the sign of exciton split Cotton effects and has been qualitatively defined: a positive exciton chirality corresponds to right-handed or clock-wise screwness. However, it is desirable to define exciton chirality in a quantitative manner.

It is easily understandable, by considering equations 10.84 through 10.92, that if the signs of interaction energy V_{ij} and triple product $\vec{R}_{ij} \cdot (\vec{\mu}_{i0a} \times \vec{\mu}_{j0a})$ are identical with each other, the first Cotton effect at longer wavelength is positive and if the signs are different from each other, the first Cotton effect is negative. Therefore, the sign of exciton chirality agrees with the sign of quadruple product $\vec{R}_{ij} \cdot (\vec{\mu}_{i0a} \times \vec{\mu}_{j0a}) V_{ij}$.

This conclusion was theoretically verified by the following calculation, which clarified that the quadruple product defines not only the sign of exciton chirality —, i.e., sign of Cotton effects — but also the amplitude of exciton Cotton effects.

In analogy with the case of binary systems discussed in the previous sections, the molecular exciton theory is applicable to UV and CD spectra of N-mer; i.e., a system with N identical chromophores. When N chromophores possessing strong $\pi \rightarrow \pi^*$ transitions $(0 \rightarrow a)$ interact with each other, the excited state a splits into N energy levels. The excitation wavenumber σ_k to the $\underline{k}$th excited level of the whole system is represented by equation 10.99,

$$\sigma_k - \sigma_0 = \sum_{i=1}^{N} \sum_{j \neq i}^{N} C_{ik} C_{jk}^{*} V_{ij} \qquad (10.99)$$

where σ_0 is the excitation wavenumber of the isolated noninteracting chromophore, C_{ik} and C_{jk}^{*} are coefficients of the corresponding $\underline{k}$th wave function, and V_{ij} is the transition dipole interaction energy between two chromophores i and j.

Similarly, the $\underline{k}$th rotational strength R^k due to the exciton coupling mechanism is

$$R^k = \pi \sigma_0 \sum_{i=1}^{N} \sum_{j \neq i}^{N} C_{ik} C_{jk}^{*} \vec{R}_j \cdot (\vec{\mu}_{j0a} \times \vec{\mu}_{i0a}) \qquad (10.100)$$

where $\vec{R}_j$ is the distance vector from the origin to chromophore j, and $\vec{\mu}_{i0a}$ and $\vec{\mu}_{j0a}$ are electric transition moment vectors of groups i and j. If we take the real wave function for the N-mer and combine the two terms of $\vec{R}_j \cdot (\vec{\mu}_{j0a} \times \vec{\mu}_{i0a})$ and $\vec{R}_i \cdot (\vec{\mu}_{i0a} \times \vec{\mu}_{j0a})$, the following origin-independent formula of rotational strength is obtained:

$$\sigma_k - \sigma_0 = 2 \sum_{i=1}^{N} \sum_{j > i}^{N} C_{ik} C_{jk} V_{ij} \qquad (10.101)$$

$$R^k = -\pi\sigma_0 \sum_{i=1}^{N} \sum_{j>i}^{N} C_{ik}C_{jk}\vec{R}_{ij} \cdot (\vec{\mu}_{i0a} \times \vec{\mu}_{j0a}) \tag{10.102}$$

where $\vec{R}_{ij}$ is the interchromophoric distance vector from i to j, and V_{ij} (expressed in cm^{-1} unit) is approximated as follows:

$$V_{ij} = \mu_{i0a}\mu_{j0a}R_{ij}^{-3}\{\vec{e}_i \cdot \vec{e}_j - 3(\vec{e}_i \cdot \vec{e}_{ij})(\vec{e}_j \cdot \vec{e}_{ij})\} \tag{10.103}$$

where $\vec{e}_i$, $\vec{e}_j$, and $\vec{e}_{ij}$ are unit vectors of $\vec{\mu}_{i0a}$, $\vec{\mu}_{j0a}$, and $\vec{R}_{ij}$, respectively.

Next, if a Gaussian distribution is approximated for the component CD Cotton effects, the curve of the kth Cotton effect is formulated as

$$\Delta\varepsilon(\sigma)^k = \Delta\varepsilon_{max}^k \exp\{-((\sigma-\sigma_k)/\Delta\sigma)^2\} \tag{10.104}$$

where $\Delta\varepsilon_{max}^k$ is the maximum value of the Cotton effect, and $\Delta\sigma$ is the standard deviation of the Gaussian distribution. On the other hand, the kth experimental rotational strength R^k is expressed by

$$R^k = 2.296 \times 10^{-39} \int_0^{\infty} \Delta\varepsilon(\sigma)^k/\sigma \, d\sigma \tag{10.105}$$

$$= (2.296 \times 10^{-39}/\sigma_k) \int_0^{\infty} \Delta\varepsilon(\sigma)^k \, d\sigma \quad \text{(cgs unit)} \tag{10.106}$$

Substitution of equation 10.104 into equation 10.106 and integration gives

$$R^k = 2.296 \times 10^{-39} \sqrt{\pi} \Delta \varepsilon_{max}{}^k \Delta \sigma / \sigma_k \tag{10.107}$$

where σ_k can be replaced by σ_0 because $\sigma_k \simeq \sigma_0$.   From   equations   10.104   and 10.107, the calculated CD curve of a whole system is derived as follows:

$$\Delta \varepsilon (\sigma) = \{\sigma_0 / (2.296 \times 10^{-39} \sqrt{\pi} \Delta \sigma)\} \sum_{k=1}^{N} R^k \exp\{-((\sigma - \sigma_k) / \Delta \sigma)^2\} \tag{10.108}$$

where $\sigma_k - \sigma_0$ and R^k are calculable by equations 10.101 and 10.102, respective-ly.

A general equation of the Taylor expansion is formulated as:

$$f(x) = \sum_{n=0}^{\infty} (f^{(n)}(a) / n!) (x-a)^n \tag{10.109}$$

According to this theorem, equation 10.108 is expanded against $\sigma_k / \Delta \sigma$ around $\sigma_0 / \Delta \sigma$.   The Taylor expansion of equation 10.108 is calculated as follows:

$$\Delta \varepsilon (\sigma) = C \sum_{k=1}^{N} R^k \exp\{-((\sigma - \sigma_k) / \Delta \sigma)^2\} \tag{10.110}$$

where $\sigma_k / \Delta \sigma$ corresponds to x and $\sigma_0 / \Delta \sigma$ corresponds to a.

$$C = \sigma_0 / (2.296 \times 10^{-39} \sqrt{\pi} \Delta \sigma) \tag{10.111}$$

The first term

$$= \sum_{k=1}^{N} \{f^{(0)} (\sigma_0/\Delta\sigma)/0!\} (\frac{\sigma_k - \sigma_0}{\Delta\sigma})^0$$

$$= \sum_{k=1}^{N} f(\sigma_0/\Delta\sigma)$$

$$= C \sum_{k=1}^{N} R^k \exp\{-(\frac{\sigma - \sigma_0}{\Delta\sigma})^2\}$$

$$= C \exp\{-(\frac{\sigma - \sigma_0}{\Delta\sigma})^2\} \sum_{k=1}^{N} R^k \qquad (10.112)$$

Thus, the first term vanishes because of the sum rule

$$\sum_{k=1}^{N} R^k = 0 \qquad (10.113)$$

The sum rule is easily understandable in a binary system because

$$\sum_{k=1}^{2} R^k = R^\alpha + R^\beta = R^\alpha - R^\alpha = 0 \qquad (10.114)$$

The second term

$$= \sum_{k=1}^{N} \{f^{(1)} (\sigma_0/\Delta\sigma)/1!\} (\frac{\sigma_k - \sigma_0}{\Delta\sigma})^1$$

$$= C \sum_{k=1}^{N} R^k \exp\{-(\frac{\sigma - \sigma_0}{\Delta\sigma})^2\} (-2) (\frac{\sigma - \sigma_0}{\Delta\sigma}) (-1) (\frac{\sigma_k - \sigma_0}{\Delta\sigma})$$

$$= 2C (\frac{\sigma - \sigma_0}{\Delta\sigma}) \exp\{-(\frac{\sigma - \sigma_0}{\Delta\sigma})^2\} \sum_{k=1}^{N} R^k (\frac{\sigma_k - \sigma_0}{\Delta\sigma}) \qquad (10.115)$$

The third term

$$= \sum_{k=1}^{N} \{f^{(2)} (\sigma_0/\Delta\sigma)/2!\} (\frac{\sigma_k - \sigma_0}{\Delta\sigma})^2$$

$$= (C/2) \sum_{k=1}^{N} R^k [2\exp\{-(\frac{\sigma - \sigma_0}{\Delta\sigma})^2\} (-2) (\frac{\sigma - \sigma_0}{\Delta\sigma}) (-1) (\frac{\sigma - \sigma_0}{\Delta\sigma})$$

$$+ 2(-1) \exp\{-(\frac{\sigma - \sigma_0}{\Delta\sigma})^2\}] (\frac{\sigma_k - \sigma_0}{\Delta\sigma})^2$$

$$= C \exp\{-(\frac{\sigma - \sigma_0}{\Delta\sigma})^2\} \{2 (\frac{\sigma - \sigma_0}{\Delta\sigma})^2 - 1\} \sum_{k=1}^{N} R^k (\frac{\sigma_k - \sigma_0}{\Delta\sigma})^2 \qquad (10.116)$$

The third and higher terms can be neglected because usually, $\sigma_k - \sigma_0 = 50 \sim 350$ cm^{-1}, $\Delta\sigma \approx 2500$ cm^{-1}, and therefore $(\sigma_k - \sigma_0)/\Delta\sigma = 0.02 \sim 0.14$.   Thus, the second term remains and the CD curve is expressed by

$$\Delta\varepsilon(\sigma) = \frac{2\sigma_0}{2.296 \times 10^{-39}\sqrt{\pi}\Delta\sigma} \exp\{-(\frac{\sigma-\sigma_0}{\Delta\sigma})^2\}(\frac{\sigma-\sigma_0}{\Delta\sigma}) \sum_{k=1}^{N} R^k (\frac{\sigma_k-\sigma_0}{\Delta\sigma}) \qquad (10.117)$$

Substitution of equations 10.101 and 10.102 into equation 10.117 gives

$$\Delta\varepsilon(\sigma) = \frac{4\sqrt{\pi}\sigma_0^2}{2.296 \times 10^{-39}\Delta\sigma^2}(\frac{\sigma_0-\sigma}{\Delta\sigma}) \exp\{-(\frac{\sigma_0-\sigma}{\Delta\sigma})^2\}$$

$$\sum_{k=1}^{N} \{ \sum_{i=1}^{N} \sum_{j>i}^{N} C_{ik}C_{jk}\vec{R}_{ij} \cdot (\vec{\mu}_{i0a} \times \vec{\mu}_{j0a}) \}\{ \sum_{i=1}^{N} \sum_{j>i}^{N} C_{ik}C_{jk}V_{ij} \} \qquad (10.118)$$

This is the expanded CD equation of N-mer; the coefficients C_{ik} and C_{jk}, which depend on the geometry of chromophores, can be obtained by solving the secular equation of the Nth order.

   In binary systems, the excited state is split into two levels, α and β, as a result of transition dipole interaction. Coefficients for α-state, $1/\sqrt{2}$ and $-1/\sqrt{2}$, and for β-state, $1/\sqrt{2}$ and $1/\sqrt{2}$, are independent of the mutual geometry of two chromophores. Therefore, equation 10.118 is further simplified as

$$\Delta\varepsilon(\sigma) = \frac{4\sqrt{\pi}\sigma_0^2}{2.296\times10^{-39}\Delta\sigma^2}(\frac{\sigma_0-\sigma}{\Delta\sigma})\exp\{-(\frac{\sigma_0-\sigma}{\Delta\sigma})^2\}$$

$$\{(1/\sqrt{2})(-1/\sqrt{2})\vec{R}_{ij}\cdot(\vec{\mu}_{i0a}\times\vec{\mu}_{j0a})(1/\sqrt{2})(-1/\sqrt{2})V_{ij}$$

$$+(1/\sqrt{2})(1/\sqrt{2})\vec{R}_{ij}\cdot(\vec{\mu}_{i0a}\times\vec{\mu}_{j0a})(1/\sqrt{2})(1/\sqrt{2})V_{ij}\}$$

$$\Delta\varepsilon(\sigma) = \frac{2\sqrt{\pi}\sigma_0^2}{2.296\times10^{-39}\Delta\sigma^2}(\frac{\sigma_0-\sigma}{\Delta\sigma})\exp\{-(\frac{\sigma_0-\sigma}{\Delta\sigma})^2\}\vec{R}_{ij}\cdot(\vec{\mu}_{i0a}\times\vec{\mu}_{j0a})V_{ij} \qquad (10.119)$$

A   B

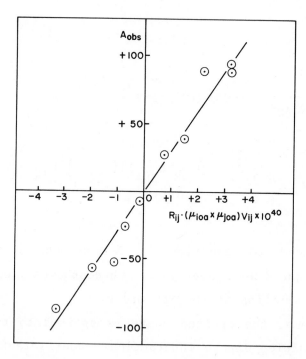

Figure 10-10. *The linear relation between* A_{obsd} *and exciton chirality* $\vec{R}_{ij}\cdot(\vec{\mu}_{i0a}\times\vec{\mu}_{j0a})V_{ij}$ *in the case of steroidal glycol bis(p-dimethylaminobenzoates) discussed in section 3-1-A. [Adapted from reference 3].*

where σ_0 and $\Delta\sigma$ are obtainable from the electronic spectrum of the corresponding chromophore.   Equation 10.119, expressing the split Cotton effects of dimers, shows that term B, $\vec{R}_{ij} \cdot (\vec{\mu}_{i0a} \times \vec{\mu}_{j0a}) V_{ij}$, which is calculable from the mutual configuration of two transition dipole moments, determines the sign and amplitude of the coupled Cotton effects.   On the other hand, term A, which has no configurational parameters, represents an anomalous dispersion curve with positive and negative extrema.

Figure 10-10 shows that a linear relation holds between A_{obsd} $(A=\Delta\varepsilon_1-\Delta\varepsilon_2)$ and the quadruple term $\vec{R}_{ij} \cdot (\vec{\mu}_{i0a} \times \vec{\mu}_{j0a}) V_{ij}$ for various steroidal bis(p-dimethylaminobenzoates) discussed in section 3-1-A. Thus, it is reasonable to take the term B, $\vec{R}_{ij} \cdot (\vec{\mu}_{i0a} \times \vec{\mu}_{j0a}) V_{ij}$, as the quantitative definition of exciton chirality in a system having two interacting and identical chromophores.

The above definition holds for non-Gaussian distribution curves; equation 10.108, can be modified as

$$\Delta\varepsilon(\sigma) = \frac{\sigma_0}{2.296\times10^{-39}\int_0^\infty f(\frac{\sigma-\sigma_0}{\Delta\sigma})d\sigma} \sum_{k=1}^{N} R^k f(\frac{\sigma-\sigma_k}{\Delta\sigma}) \qquad (10.120)$$

where $f[(\sigma-\sigma_0)/\Delta\sigma]$ is the function describing the shape of actual UV ($f(0)=1.0$) and $\Delta\sigma$ is an arbitrary value. When $f[(\sigma-\sigma_0)/\Delta\sigma]$ is approximated by the Gaussian distribution, $\Delta\sigma$ is the standard deviation.   The Taylor expansion of equation 10.120 gives equation 10.121 as the second term:

$$\Delta\varepsilon(\sigma) = \frac{-\sigma_0 f^{(1)}(\frac{\sigma-\sigma_0}{\Delta\sigma})}{2.296\times10^{-39}\int_0^\infty f(\frac{\sigma-\sigma_0}{\Delta\sigma})d\sigma} \sum_{k=1}^N R^k(\frac{\sigma_k-\sigma_0}{\Delta\sigma})$$

(10.121)

where $f^{(1)}[(\sigma-\sigma_0)/\Delta\sigma]$ indicates the value of $\frac{d}{dx}f(x)$ when $x=(\sigma-\sigma_0)/\Delta\sigma$. The first term of the expansion is zero, and other higher terms are negligible because of $(\sigma_k-\sigma_0)/\Delta\sigma \ll 1$. In addition, in the case of binary systems (N=2), all of the odd terms vanish on the basis of the sum rule ($\sum_{k=1}^N R^k=0$). Since the values R^k and $\sigma_k-\sigma_0$ are expressed by equations 10.102 and 10.101, the exciton chirality of binary system is quantitatively defined by the quadruple product, $\vec{R}_{ij}\cdot(\vec{\mu}_{i0a}\times\vec{\mu}_{j0a})V_{ij}$.

The previous qualitative definition of exciton chirality that a clockwise or right-handed twist gives a positive split Cotton effect at longer wavelength is in agreement with the present quantitative definition. These aspects make the CD exciton chirality method a simple and nonambiguous means for determining absolute configurations.

The term B can be modified as follows:

$$\text{term B} = \vec{R}_{ij}\cdot(\vec{\mu}_{i0a}\times\vec{\mu}_{j0a})V_{ij}$$

$$= D_{i0a}D_{j0a}R_{ij}^{-2}\vec{e}_{ij}\cdot(\vec{e}_i\times\vec{e}_j)\{\vec{e}_i\cdot\vec{e}_j-3(\vec{e}_i\cdot\vec{e}_{ij})(\vec{e}_j\cdot\vec{e}_{ij})\}$$

(10.122)

where D_{i0a} and D_{j0a} are transition dipole strengths of groups i and j, respectively. This equation indicates that *the Cotton effect amplitude is inversely proportional to the square of interchromophoric distance* provided the remaining angular part is the same.

In the next chapter, numerical calculation of exciton CD effects of actual various compounds is discussed by employing the present theoretical results.

10-9. Circular Dichroism of Nondegenerate Systems

Exciton split Cotton effects are observed also in the binary system composed of two different chromophores having similar λ_{max} positions. In fact, as exemplified in Chapters 5 and 7, interaction between the benzoate and polyacene or enone groups generates split Cotton effects, which enabled one to determine the absolute configurations of the systems. In this section, circular dichroism of nondegenerate systems is theoretically discussed.

For a nondegenerate system in which groups i and j undergo $0 \rightarrow a$ and $0 \rightarrow b$ transitions, respectively, as shown in Figure 10-11, the following wavefunctions and energy levels are obtained by solving the secular equation 10.123:

$$\begin{vmatrix} \sigma_a - E & V_{ij} \\ V_{ij} & \sigma_b - E \end{vmatrix} = 0 \tag{10.123}$$

$$E^\alpha = (1/2)(\sigma_a + \sigma_b) - (1/2)\{(\sigma_b - \sigma_a)^2 + 4V_{ij}^2\}^{0.5} \tag{10.124}$$

$$\psi^\alpha = C_i^\alpha \phi_{ia} \phi_{j0} + C_j^\alpha \phi_{i0} \phi_{jb} \tag{10.125}$$

$$E^{\beta} = (1/2)(\sigma_a + \sigma_b) + (1/2)\{(\sigma_b - \sigma_a)^2 + 4V_{ij}^2\}^{0.5} \qquad (10.126)$$

$$\psi^{\beta} = c_i^{\beta}\phi_{ia}\phi_{j0} + c_j^{\beta}\phi_{i0}\phi_{jb} \qquad (10.127)$$

For ground state,

$$\psi^0 = \phi_{i0}\phi_{j0} \qquad (10.128)$$

Employing these equations, electric and magnetic transition moments of α and β transitions are calculated as follows:

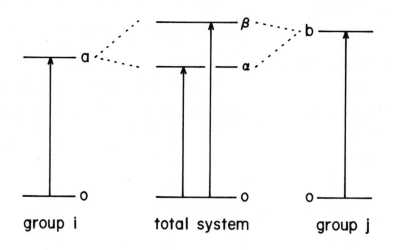

Figure 10-11. *Interaction and energy levels of nondegenerate system in which groups i and j undergo 0→a and 0→b transitions, respectively.*

Electric transition moments.

$$\langle 0|\vec{\mu}|\alpha\rangle = \int \psi^{0}\vec{\mu}\psi^{\alpha}d\tau$$

$$= \int \phi_{i0}\phi_{j0}\Sigma\vec{\mu}\{c_i{}^{\alpha}\phi_{ia}\phi_{j0}+c_j{}^{\alpha}\phi_{i0}\phi_{jb}\}d\tau$$

$$= c_i{}^{\alpha}\int \phi_{i0}\vec{\mu}_i\phi_{ia}d\tau_i\int \phi_{j0}\phi_{j0}d\tau_j$$

$$+c_i{}^{\alpha}\int \phi_{i0}\phi_{ia}d\tau_i\int \phi_{j0}\vec{\mu}_j\phi_{j0}d\tau_j$$

$$+c_j{}^{\alpha}\int \phi_{j0}\vec{\mu}_j\phi_{jb}d\tau_j$$

$$\langle 0|\vec{\mu}|\alpha\rangle = c_i{}^{\alpha}\vec{\mu}_{i0a}+c_j{}^{\alpha}\vec{\mu}_{j0b} \qquad\qquad (10.129)$$

where

$$\vec{\mu}_{i0a} = \int \phi_{i0}\vec{\mu}_i\phi_{ia}d\tau_i \qquad\qquad (10.130)$$

$$\vec{\mu}_{j0b} = \int \phi_{j0}\vec{\mu}_j\phi_{jb}d\tau_j \qquad\qquad (10.131)$$

Similarly,

$$\langle 0|\vec{\mu}|\beta\rangle = c_i{}^{\beta}\vec{\mu}_{i0a}+c_j{}^{\beta}\vec{\mu}_{j0b} \qquad\qquad (10.132)$$

Magnetic transition moments.

$$\langle \alpha | \vec{M} | 0 \rangle = \int \psi^{\alpha} \vec{M} \psi^{0} d\tau$$

$$= \int (C_i^{\alpha} \phi_{ia} \phi_{j0} + C_j^{\alpha} \phi_{i0} \phi_{jb})(e/2mc) \sum \vec{r} \times \vec{p} (\phi_{i0} \phi_{j0}) d\tau$$

$$= C_i^{\alpha} \{ (e/2mc) \vec{R}_i \times \vec{p}_{ia0} + \vec{m}_{ia0} \}$$

$$+ C_j^{\alpha} \{ (e/2mc) \vec{R}_j \times \vec{p}_{jb0} + \vec{m}_{jb0} \} \qquad (10.133)$$

where

$$\vec{p}_{jb0} = \int \phi_{jb} \vec{p}_j \phi_{j0} d\tau_j \qquad (10.134)$$

$$\vec{m}_{jb0} = \int \phi_{jb} \vec{m}_j \phi_{j0} d\tau_j \qquad (10.135)$$

Substitution of the equation

$$\vec{p}_{ib0} = (2\pi imc/e) \sigma_b \vec{\mu}_{i0b} \qquad (10.136)$$

into equation 10.133 gives

$$\langle \alpha | \vec{M} | 0 \rangle = C_i^{\alpha} \{ i\pi\sigma_a \vec{R}_i \times \vec{\mu}_{i0a} + \vec{m}_{ia0} \}$$

$$+ C_j^{\alpha} \{ i\pi\sigma_b \vec{R}_j \times \vec{\mu}_{j0b} + \vec{m}_{jb0} \} \qquad (10.137)$$

Similarly,

$$\langle\beta|\vec{M}|0\rangle = C_i^{\beta}\{i\pi\sigma_a\vec{R}_i\times\vec{\mu}_{i0a}+\vec{m}_{ia0}\}$$

$$+C_j^{\beta}\{i\pi\sigma_b\vec{R}_j\times\vec{\mu}_{j0b}+\vec{m}_{jb0}\} \tag{10.138}$$

Rotational strength is formulated as follows using equations 10.129 and 10.137:

$$R^{\alpha} = \mathrm{Im}\{\langle 0|\vec{\mu}|\alpha\rangle\cdot\langle\alpha|\vec{M}|0\rangle\}$$

$$= \mathrm{Im}[\,(C_i^{\alpha}\vec{\mu}_{i0a}+C_j^{\alpha}\vec{\mu}_{j0b})\cdot\{C_i^{\alpha}(i\pi\sigma_a\vec{R}_i\times\vec{\mu}_{i0a}+\vec{m}_{ia0})$$

$$+C_j^{\alpha}(i\pi\sigma_b\vec{R}_j\times\vec{\mu}_{j0b}+\vec{m}_{jb0})\}\,]$$

$$= \mathrm{Im}\{\,(C_i^{\alpha}\vec{\mu}_{i0a}+C_j^{\alpha}\vec{\mu}_{j0b})\cdot(C_i^{\alpha}\vec{m}_{ia0}+C_j^{\alpha}\vec{m}_{jb0})\}$$

$$+C_i^{\alpha}C_j^{\alpha}\pi\sigma_a\vec{R}_i\cdot(\vec{\mu}_{i0a}\times\vec{\mu}_{j0b})-C_i^{\alpha}C_j^{\alpha}\pi\sigma_b\vec{R}_j\cdot(\vec{\mu}_{i0a}\times\vec{\mu}_{j0b}) \tag{10.139}$$

In order to combine the second and third terms, σ_a and σ_b are approximated by the geometric average $\sqrt{\sigma_a\sigma_b}$.

$$R^{\alpha} = \mathrm{Im}\{\,(C_i^{\alpha}\vec{\mu}_{i0a}+C_j^{\alpha}\vec{\mu}_{j0b})\cdot(C_i^{\alpha}\vec{m}_{ia0}+C_j^{\alpha}\vec{m}_{jb0})\}$$

$$-C_i^{\alpha}C_j^{\alpha}\pi\sqrt{\sigma_a\sigma_b}\vec{R}_{ij}\cdot(\vec{\mu}_{i0a}\times\vec{\mu}_{j0b}) \tag{10.140}$$

Similarly,

$$R^\beta = \text{Im}\{ (C_i^\beta \vec{\mu}_{i0a} + C_j^\beta \vec{\mu}_{j0b}) \cdot (C_i^\beta \vec{m}_{ia0} + C_j^\beta \vec{m}_{jb0}) \}$$

$$-C_i^\beta C_j^\beta \pi \sqrt{\sigma_a \sigma_b} \vec{R}_{ij} \cdot (\vec{\mu}_{i0a} \times \vec{\mu}_{j0b}) \qquad\qquad (10.141)$$

Since we are dealing with $\pi \to \pi^*$ transitions of planar chromophores, the first term can be neglected as in the case of degenerate systems.

Next, coefficients C_i^α, C_j^α, C_i^β, and C_j^β are calculated from the following equations:

secular equation,

$$(\sigma_a - E^\alpha) C_i^\alpha + V_{ij} C_j^\alpha = 0 \qquad\qquad (10.142)$$

normalization condition,

$$(C_i^\alpha)^2 + (C_j^\alpha)^2 = 1 \qquad\qquad (10.143)$$

$$C_i^\alpha = V_{ij} / \{ V_{ij}^2 + (\sigma_a - E^\alpha)^2 \}^{0.5} \qquad\qquad (10.144)$$

$$C_j^\alpha = -(\sigma_a - E^\alpha) / \{ V_{ij}^2 + (\sigma_a - E^\alpha)^2 \}^{0.5} \qquad\qquad (10.145)$$

Therefore, product $C_i^\alpha C_j^\alpha$ is formulated as:

$$C_i^\alpha C_j^\alpha = -V_{ij} (\sigma_a - E^\alpha) / \{ V_{ij}^2 + (\sigma_a - E^\alpha)^2 \} \qquad\qquad (10.146)$$

Similarly,

$$c_i^{\beta} c_j^{\beta} = -V_{ij}(\sigma_a - E^{\beta})/\{V_{ij}^2 + (\sigma_a - E^{\beta})^2\}$$

(10.147)

Now, parameters t and δ are introduced:

$$\delta = \sigma_b - \sigma_a$$

(10.148)

$$t = \{(\sigma_b - \sigma_a)^2 + 4V_{ij}^2\}^{0.5}$$

$$= (\delta^2 + 4V_{ij}^2)^{0.5}$$

(10.149)

Therefore,

$$4V_{ij}^2 = t^2 - \delta^2$$

(10.150)

$$\sigma_a - E^{\alpha} = (t - \delta)/2$$

(10.151)

$$\sigma_a - E^{\beta} = -(t + \delta)/2$$

(10.152)

Substitution of these equations into equations 10.146 and 10.147 gives:

$$c_i^{\alpha} c_j^{\alpha} = -V_{ij}/t$$

(10.153)

$$c_i^{\beta} c_j^{\beta} = V_{ij}/t$$

(10.154)

Consequently, wavefunctions, energy values, and rotational strengths are summarized as follows:

α-state

$$\psi^\alpha = C_i{}^\alpha \phi_{ia}\phi_{j0} + C_j{}^\alpha \phi_{i0}\phi_{jb} \tag{10.155}$$

$$E^\alpha = (1/2)(\sigma_a + \sigma_b) - (1/2)\{(\sigma_b - \sigma_a)^2 + 4V_{ij}{}^2\}^{0.5} \tag{10.156}$$

$$R^\alpha = [\pi V_{ij}\sqrt{\sigma_a\sigma_b}/\{(\sigma_b - \sigma_a)^2 + 4V_{ij}{}^2\}^{0.5}]\vec{R}_{ij} \cdot (\vec{\mu}_{i0a} \times \vec{\mu}_{j0b}) \tag{10.157}$$

β-state

$$\psi^\beta = C_i{}^\beta \phi_{ia}\phi_{j0} + C_j{}^\beta \phi_{i0}\phi_{jb} \tag{10.158}$$

$$E^\beta = (1/2)(\sigma_a + \sigma_b) + (1/2)\{(\sigma_b - \sigma_a)^2 + 4V_{ij}{}^2\}^{0.5} \tag{10.159}$$

$$R^\beta = -[\pi V_{ij}\sqrt{\sigma_a\sigma_b}/\{(\sigma_b - \sigma_a)^2 + 4V_{ij}{}^2\}^{0.5}]\vec{R}_{ij} \cdot (\vec{\mu}_{i0a} \times \vec{\mu}_{j0b}) \tag{10.160}$$

Thus, rotational strengths of α and β transitions are same in amplitude but of opposite signs. Namely, sum rule ($\sum R = R^\alpha + R^\beta = 0$) holds also for nondegenerate systems.

As indicated in equations 10.157 and 10.160, if the quadruple product $\vec{R}_{ij} \cdot (\vec{\mu}_{i0a} \times \vec{\mu}_{j0b})V_{ij}$ is positive, the first Cotton effect at longer wavelengths is positive and the second one at shorter wavelengths is negative. Thus, the present theoretical calculations indicate that the exciton chirality method is applicable to nondegenerate systems.

Numerical calculations and experimental data of nondegenerate steroidal dibenzoates are discussed in section 3-1-D.

If interaction energy V_{ij} is small and negligible in comparison with $\sigma_b - \sigma_a$, the equation of rotational strength is simplified:

$$R^{\alpha} = \{\pi V_{ij}\sqrt{\sigma_a \sigma_b}/(\sigma_b - \sigma_a)\}\vec{R}_{ij} \cdot (\vec{\mu}_{i0a} \times \vec{\mu}_{j0b})$$

$$= \{\pi V_{ij}\sqrt{\sigma_a \sigma_b}(\sigma_b + \sigma_a)/(\sigma_b^2 - \sigma_a^2)\}\vec{R}_{ij} \cdot (\vec{\mu}_{i0a} \times \vec{\mu}_{j0b}) \qquad (10.161)$$

By using the approximation of $\sigma_b + \sigma_a \simeq 2\sqrt{\sigma_a \sigma_b}$

$$R^{\alpha} = \{2\pi \sigma_a \sigma_b V_{ij}/(\sigma_b^2 - \sigma_a^2)\}\vec{R}_{ij} \cdot (\vec{\mu}_{i0a} \times \vec{\mu}_{j0b}) \qquad (10.162)$$

This equation is identical with that derived by Tinoco[10,16] for studying the chiroptical properties of biopolymers.

References

1. N. Harada, S. Suzuki, H. Uda, and K. Nakanishi, J. Am. Chem. Soc. 93, 5577 (1971).

2. N. Harada, J. Am. Chem. Soc. 95, 240 (1973).

3. N. Harada, S. -M. L. Chen, and K. Nakanishi, J. Am. Chem. Soc. 97, 5345 (1975).

4. N. Harada, Y. Takuma, H. Uda, J. Am. Chem. Soc. 100, 4029 (1978).

5. N. Harada, unpublished data.

6. V. L. Rosenfeld, Z. Phys. 52, 161 (1928).

7. S. -M. L. Chen, N. Harada, and K. Nakanishi, J. Am. Chem. Soc. 96, 7352 (1974).

8. A. S. Davydov, Theory of Molecular Excitons Trans. M. Kasha and M. Oppenheimer, Jr. (New York: McGraw-Hill, 1962).

9. A. S. Davydov, Zhur. Eksptl. i Teoret. Fiz. 18, 210 (1948). [C.A. 43, 4575f (1949)]

10. I. Tinoco, Jr., Advan. Chem. Phys. 4, 113 (1962).

11. H. Eyring, J. Walter, and G. E. Kimball, <u>Quantum Chemistry</u> (New York:
     Wiley, 1944), Chapter 17.

12. D. Bohm, <u>Quantum Theory</u> (New York: Prentice-Hall, 1951), p. 427.

13. H. Eyring, J. Walter, and G. E. Kimball, <u>Quantum Chemistry</u> (New York:
     Wiley, 1944), Chapter 3.

14. N. Harada and K. Nakanishi, <u>Acc. Chem. Res.</u> <u>5</u>, 257 (1972).

15. G. Wagniere and W. Hug, <u>Tetrahedron Lett.</u> 1970, p. 4765. W. Hug and
     G. Wagniere, <u>Tetrahedron</u>, <u>28</u>, 1241 (1972).

16. I. Tinoco, Jr., in <u>Fundamental Aspects and Recent Developments in Optical
     Rotary Dispersion and Circular Dichroism</u> ed. F. Ciardelli and P. Salva-
     dori (London: Heyden, 1973), Chapter 2.4.

XI. NUMERICAL CALCULATION OF EXCITON CD COTTON EFFECTS BY THE EXCITON CHIRALITY METHOD

11-1. Numerical Calculation of CD Spectra of Bis(p-chlorobenzoates)[1]

In this chapter, the theoretical results discussed in the preceding chapter are applied to numerical calculations of CD spectra of various compounds. The first case is cholest-5-ene-3β,4β-diol bis(p-chlorobenzoate).

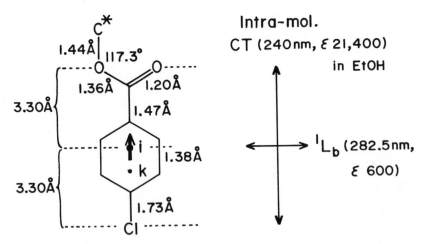

Figure 11-1. Location and direction of the transition point dipole in p-chlorobenzoate chromophore.

It is first necessary to evaluate the position of a transition dipole moment in the geometry of the p-chlorobenzoate chromophore.   As shown in Figure 11-1, geometric parameters were taken from X-ray crystallographic data of pertinent compounds; the angle of the ether oxygen is 117.3° while the remaining angles of sp^2 hybridization are 120°.

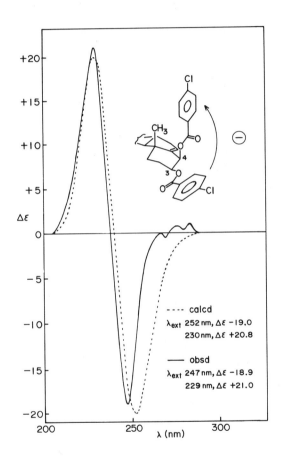

+20

+15

+10

+5

Δε

0

−5

−10

---- calcd
λ_{ext} 252 nm, Δε −19.0
230 nm, Δε +20.8

—— obsd
λ_{ext} 247 nm, Δε −18.9
229 nm, Δε +21.0

−15

−20

200   250   300

λ (nm)

Figure 11-2.   *Comparison of calculated (dotted line) and observed (solid line) CD curves of cholest-5-ene- 3β,4β-diol bis(p-chlorobenzoate).* [*Adapted from reference 1.*]

The intramolecular charge transfer transition of p-chlorobenzoate at 240 nm (ε 21,400) is polarized along the long axis of the chromophore.   Therefore, it is assumed that the transition dipole moment is located at the mid-point of a line connecting chlorine atom and two oxygen atoms, as illustrated in Figure 11-1.   In order to define the direction of the dipole moment, an additional point is needed and this is named point k in Figure 11-1.   Point k, which lies on the long axis of the p-chlorobenzoate chromophore, is used only for defining the direction of transition moment.

The conformational geometry of a vicinal dibenzoate moiety is depicted in Figure 11-2; each benzoyloxy group adopts the most stable staggered conformation.

The Cartesian coordinate system of the present dibenzoate moiety was chosen as shown in Figure 11-3, where x-axis is the C(3)-C(4) bond, z-axis is

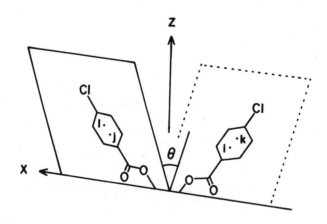

Figure 11-3.   *The Cartesian coordinates and dihedral angle θ of vicinal bis(p-chlorobenzoate) with negative exciton chirality.*

the C_2-symmetrical axis of the dibenzoate moiety, and the dihedral angle θ between two benzoate planes is 60°. The Cartesian coordinates of the transition dipole moment and the additional point are readily calculable as follows.

The local axis z' was placed so that plane xz' contains one of two benzoate chromophores; the local coordinates of point dipole and additional point contained in the plane xz' were then calculated. The principal coordinates were estimated as:

$$x = a \qquad\qquad\qquad (11.1)$$
$$y = z'\sin(\theta/2) = z'\sin 30° = b\sin 30° \qquad (11.2)$$
$$z = z'\cos(\theta/2) = z'\cos 30° = b\cos 30° \qquad (11.3)$$

The Cartesian coordinate of the other benzoate moiety was calculable in the same way as follows:

$$x = -a \qquad\qquad\qquad (11.4)$$
$$y = -b\sin 30° \qquad\qquad\qquad (11.5)$$
$$z = b\cos 30° \qquad\qquad\qquad (11.6)$$

Now, we have the Cartesian coordinates of four points i, j, k, l to define the configuration of the two dipole moments in the dibenzoate moiety, as shown in Figure 11-3, where i and j are position of transition moments i and j, respectively, and k and l are additional points to define the direction of moments i and j, respectively.

By usage of Cartesian coordinates, we can now calculate interaction energy V_{ij} as follows:

$$\vec{e}_i \cdot \vec{e}_j = \{(x_k - x_i)(x_l - x_j) + (y_k - y_i)(y_l - y_j) + (z_k - z_i)(z_l - z_j)\}/(R_{ik}R_{jl}) \qquad (11.7)$$

where

$$R_{ik} = \{(x_i - x_k)^2 + (y_i - y_k)^2 + (z_i - z_k)^2\}^{0.5} \qquad (11.8)$$

$$R_{jl} = \{(x_j - x_l)^2 + (y_j - y_l)^2 + (z_j - z_l)^2\}^{0.5} \qquad (11.9)$$

$$\vec{e}_i \cdot \vec{e}_{ij} = \{(x_k - x_i)(x_j - x_i) + (y_k - y_i)(y_j - y_i) + (z_k - z_i)(z_j - z_i)\}/(R_{ik}R_{ij}) \qquad (11.10)$$

where

$$R_{ij} = \{(x_i - x_j)^2 + (y_i - y_j)^2 + (z_i - z_j)^2\}^{0.5} \qquad (11.11)$$

$$\vec{e}_j \cdot \vec{e}_{ij} = \{(x_l - x_j)(x_j - x_i) + (y_l - y_j)(y_j - y_i) + (z_l - z_j)(z_j - z_i)\}/(R_{jl}R_{ij}) \qquad (11.12)$$

Interaction energy V_{ij} is calculated as

$$V_{ij} = 5.0340 \times (4.8032)^2 \times 1000 \times r^2 \{\vec{e}_i \cdot \vec{e}_j - 3(\vec{e}_i \cdot \vec{e}_{ij})(\vec{e}_j \cdot \vec{e}_{ij})\}/R_{ij}^{3} \qquad (11.13)$$

where V_{ij} is expressed in cm^{-1} unit, r expressed in Å unit is the transition length value of the chromophores, and R_{ij} is the interchromophoric distance expressed in Å unit.

The conversion factor and physical constant employed are

$$1 \text{ erg} = 5.0340 \times 10^{15} \text{ cm}^{-1} \tag{11.14}$$

$$\text{charge of electron } e = 4.8032 \times 10^{-10} \text{ esu} \tag{11.15}$$

In the case of cholest-5-ene-3β,4β-diol bis(<u>p</u>-chlorobenzoate), transition length r is 0.992 Å and interaction energy V_{ij} is calculated to be +239.1 cm^{-1}. The positive sign of V_{ij} value indicates that the α-state is lower in energy level than the β-state; in other words, α-state corresponds to the first Cotton effect at longer wavelengths and β-state corresponds to the second Cotton effect.

The computation of rotational strength was performed as follows. In general, scalar triple product $\vec{A} \cdot (\vec{B} \times \vec{C})$ is expressed by the following determinant.

$$\vec{A} \cdot (\vec{B} \times \vec{C}) = \begin{vmatrix} A_x & A_y & A_z \\ B_x & B_y & B_z \\ C_x & C_y & C_z \end{vmatrix} \tag{11.16}$$

where A_x, A_y, and A_z are x-, y-, and z-components of vector $\vec{A}$, respectively. Therefore, triple product $\vec{e}_{ij} \cdot (\vec{e}_i \times \vec{e}_j)$ is formulated as

$$\vec{e}_{ij} \cdot (\vec{e}_i \times \vec{e}_j) = \{(x_j - x_i)(y_k - y_i)(z_1 - z_j) + (x_k - x_i)(y_1 - y_j)(z_j - z_i)$$

$$+ (x_1 - x_j)(y_j - y_i)(z_k - z_i) - (x_1 - x_j)(y_k - y_i)(z_j - z_i)$$

$$- (x_j - x_i)(y_1 - y_j)(z_k - z_i) - (x_k - x_i)(y_j - y_i)(z_1 - z_j)\}$$

$$/(R_{ij} R_{ik} R_{j1}) \tag{11.17}$$

The rotational strength R^α of α-state is calculated as

$$R^\alpha = +(1/2)\pi\sigma_0 R_{ij}\mu_{i0a}\mu_{j0a}\vec{e}_{ij} \cdot (\vec{e}_i \times \vec{e}_j)$$

$$= (1/2)\pi\sigma_0 R_{ij} \times (4.8032)^2 \times r^2 \times 10^{-4} \times \vec{e}_{ij} \cdot (\vec{e}_i \times \vec{e}_j) \tag{11.18}$$

where σ_0 is the excitation wavenumber in cm^{-1} unit, R_{ij} is the interchromo-phoric distance in Å unit, and r is in the transition length in Å unit. In the case of p-chlorobenzoate, parameters are

$$\sigma_0 = 41,666.6 \text{ cm}^{-1} \tag{11.19}$$

$$r = 0.992 \text{ Å} \tag{11.20}$$

and numerical computation gave

$$R^\alpha = -3.376 \times 10^{-38} \qquad \text{cgs unit} \tag{11.21}$$

The rotational strength R^β of β-state is necessarily

$$R^\beta = +3.376 \times 10^{-38} \text{ cgs unit} \tag{11.22}$$

Thus, the calculation results give negative first and positive second Cotton effects.

Next, we can calculate CD curve using the above data and equations 10.104 and 10.107. Substitution of parameter values R, $\Delta\sigma$, and σ_0 into equation 10.107 gives $\Delta\varepsilon_{max}$ value; in the case of p-chlorobenzoate, standard deviation $\Delta\sigma$ of the Gaussian distribution obtained from the UV spectrum is

$$\Delta\sigma = 2665.0 \text{ cm}^{-1} \tag{11.23}$$

Then, the calculated $\Delta\varepsilon_{max}$ value of α-state is

$$\Delta\varepsilon_{max}^{\ \alpha} = -129.8 \tag{11.24}$$

$$\Delta\varepsilon_{max}^{\ \beta} = +129.8 \tag{11.25}$$

These two Cotton effects are separated by a Davydov splitting of $2V_{ij} = 478.2 \text{ cm}^{-1}$; namely, sign, amplitude, and position of two Cotton effects are as follows.

$$\alpha\text{-state:} \quad \Delta\varepsilon(\sigma)^\alpha = -129.8 \exp\{-((\sigma-\sigma_\alpha)/\Delta\sigma)^2\} \tag{11.26}$$

where $\sigma_\alpha = \sigma_0 - V_{ij} = 41,666.7 - 239.1 = 41427.6 \text{ cm}^{-1}$.

β-state: $\Delta\varepsilon(\sigma)^{\beta} = +129.8 \exp\{-((\sigma-\sigma_{\beta})/\Delta\sigma)^2\}$ (11.27)

where $\sigma_{\beta} = \sigma_0 + V_{ij} = 41{,}666.7 + 239.1 = 41905.8 \text{ cm}^{-1}$.

Plots of these two CD Cotton effects are shown in Figure 11-4.

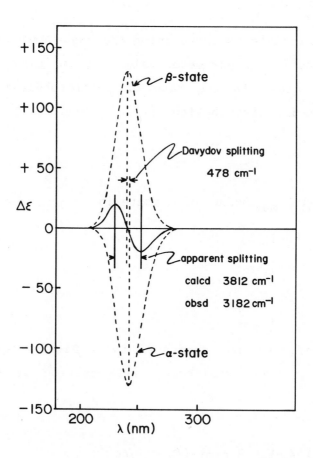

Figure 11-4.' *Theoretical component CD curves (dotted line) and its summation curve (solid line) of cholest-5-ene-3β,4β-diol bis(p-chlorobenzoate). [Adapted from reference 2.]*

Summation of the two component CD curves results in split type CD Cotton effects, as illustrated in Figure 11-4.

$$\Delta\varepsilon(\sigma) = \Delta\varepsilon(\sigma)^{\alpha} + \Delta\varepsilon(\sigma)^{\beta} \qquad (11.28)$$

The calculated Cotton effects

calcd λ_{ext} 252 nm, $\Delta\varepsilon$ −19.0
                    230 nm, $\Delta\varepsilon$ +20.8

are in excellent agreement, both in position and in intensity, with the experimental curve as illustrated in Figure 11-2.

obsd λ_{ext} 247 nm, $\Delta\varepsilon$ −18.9
                    229 nm, $\Delta\varepsilon$ +21.0

11-2. Angular Dependence of Exciton Cotton Effects of Dibenzoates[3]

The sign and amplitude of exciton split Cotton effects are functions of the dihedral angle between two benzoate chromophores. As qualitatively discussed in Chapter 3, in the case of vicinal dibenzoates, the sign of split Cotton effects remains unchanged in the range of a dihedral angle from 0° to 180°. Namely, the qualitative definition of exciton chirality holds for vicinal dibenzoates having dihedral angle in the range of 0°—180°.

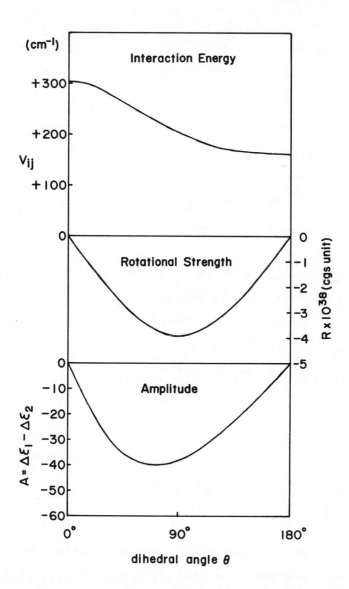

<u>*Figure 11-5.*</u>   *Angular dependence of interaction energy, rotational strength(R^{α}), and amplitude of CD Cotton effects of vicinal glycol bis(p-chlorobenzoate) with negative exciton chirality, where θ is dihedral angle between two benzoate planes.*

Table 11-1. Angular Dependence of Interaction Energy, Rotational Strength, and Amplitude of the Split Cotton Effects of Vicinal Bis(p-chlorobenzoate).

dihedral angle θ	$\vec{e}_i \cdot \vec{e}_j$	$-3(\vec{e}_i \cdot \vec{e}_{ij})(\vec{e}_j \cdot \vec{e}_{ij})$	V_{ij} (cm^{-1})	rotational strength $R^\alpha \times 10^{38}$ cgs unit	amplitude A
0°	+0.500	+0.750	+302.0	0.0	0.0
30°	+0.399	+0.872	+281.1	-1.94	-26.9
60°	+0.125	+1.220	+239.0	-3.37	-39.7
90°	-0.250	+1.719	+202.2	-3.89	-38.8
120°	-0.625	+2.235	+178.3	-3.37	-29.7
150°	-0.900	+2.620	+165.4	-1.94	-15.9
180°	-1.000	+2.762	+161.3	0.0	0.0

Now, we can ascertain the above conclusion by quantitative calculation. When the dihedral angle θ between the two benzoate planes is gradually changed from $0°$ to $180°$ (Figure 11-3), calculations of interaction energy V_{ij}, rotational strength R, and amplitude of split Cotton effects gives results shown in Table 11-1 and Figure 11-5.

In the calculation of interaction energy V_{ij}, the second term $-3(\vec{e}_i \cdot \vec{e}_{ij})$ $(\vec{e}_j \cdot \vec{e}_{ij})$ is always positive and larger than the first term $(\vec{e}_i \cdot \vec{e}_{ij})$ which changes its sign at about $70°$ (Table 11-1). Therefore, interaction energy V_{ij} which is proportional to the sum of the two terms is positive in the range of $0°$ to $180°$ (Figure 11-5).

On the other hand, the value of rotational strength R^α of α-state is negative and changes as a sine curve.

In general, rotational strength of a system with C_2-symmetry changes as a sine function against dihedral angle θ. This is easily proven by the following calculation.

The geometrical term $\vec{R}_{ij} \cdot (\vec{e}_i \times \vec{e}_j)$ of rotational strength is formulated as

$$\vec{R}_{ij} \cdot (\vec{e}_i \times \vec{e}_j) = \begin{vmatrix} x_j - x_i & y_j - y_i & z_j - z_i \\ x_k - x_i & y_k - y_i & z_k - z_i \\ x_l - x_j & y_l - y_j & z_l - z_j \end{vmatrix} / (R_{ik} R_{jl}) \qquad (11.29)$$

The Cartesian coordinates of points i, j, k, and l are expressed as

$$x_i = a, \qquad y_i = b \sin(\theta/2), \qquad z_i = b \cos(\theta/2) \tag{11.30}$$

$$x_j = -a, \qquad y_j = -b \sin(\theta/2), \qquad z_j = b \cos(\theta/2) \tag{11.31}$$

$$x_k = a', \qquad y_k = b' \sin(\theta/2), \qquad z_k = b' \cos(\theta/2) \tag{11.32}$$

$$x_l = -a', \qquad y_l = -b' \sin(\theta/2), \qquad z_l = b' \cos(\theta/2) \tag{11.33}$$

Substitution of these coordinates gives

$$\vec{R}_{ij} \cdot (\vec{e}_i \times \vec{e}_j) = \frac{\sin(\theta/2)\cos(\theta/2)}{R_{ik}R_{jl}} \begin{vmatrix} -2a & -2b & 0 \\ a'-a & b'-b & b'-b \\ -(a'-a) & -(b'-b) & b'-b \end{vmatrix}$$

$$= \frac{(1/2)\sin(\theta)}{R_{ik}R_{jl}} \begin{vmatrix} -2a & -2b & 0 \\ a'-a & b'-b & b'-b \\ 0 & 0 & 2(b'-b) \end{vmatrix}$$

$$= \sin(\theta) \times \text{constant term} \tag{11.34}$$

Thus, rotational strength R changes as a sine curve of dihedral angle θ.

11-3.   Distant Effect in the Exciton Chirality Method and Electro- Vibrational Structure of Exciton Split Cotton Effects[3]

In the following systematic studies on the distance dependence of the split Cotton effect $\Delta\varepsilon$ values, the steroidal compounds were chosen for reasons that the skeleton is rigid and that a number of simple diols were available (Table 11-2). All diols carry one of the hydroxyl functions at C-3 (Figure 11-6). As a model for the distant 1,8-glycol, a 3β,15β-diol having a six-membered D-homo ring was chosen. This was to avoid complications arising from a five-

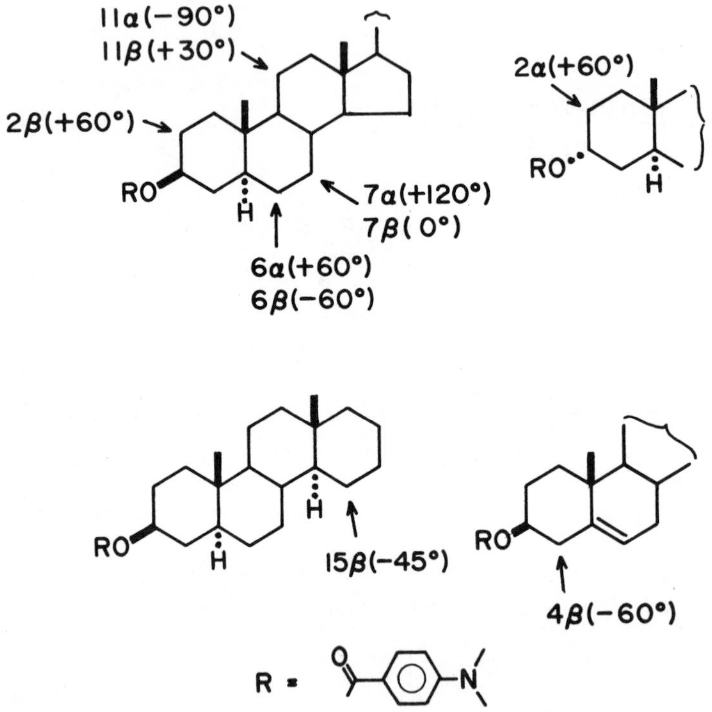

Figure 11-6. Numerals in parenthesis denote dihedral angles between two C-O bonds of diols. [Adapted from reference 3.]

membered ring, which is flexible and has bond lengths and angles differing
from six-membered rings.

The p-dimethylaminobenzoate group was chosen as the chromophore because
the intense intramolecular charge transfer band along the long axis at 309 nm
leads to strong coupled Cotton effects. Furthermore, the long wavelength location of the band at 309 nm facilitates theoretical treatments; it presumably overlaps with the short axis ^1L_b transition but contribution of the
latter has been neglected because of its low UV intensity.

In the present calculation, we have assumed that the point dipole is
located at the midpoint between the N and O-O distance as shown in Figure
11-7. The coordinates for the point dipole were computed by assuming that the

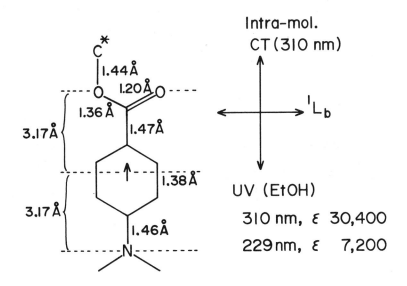

Figure 11-7.   *Location and direction of the transition point dipole in
p-dimethylaminobenzoate chromophore. [Adapted from reference 3].*

steroid skeleton adopts an ideal chair conformation, and that each benzoyloxy group adopts the most stable staggered conformation. For the benzoate group adjacent to ring junctions, only one conformation, in which it is trans to the ring junction C-H, was considered (III in Figure 11-8). However, in the case of the 3β-benzoate, which is flanked by two CH_2 moieties, two staggered conformers I and II were considered (Figure 11-8) and the arithmetic average of the two sets of Δε values (e.g., Figure 3-7, curves calcd I and calcd II) was calculated for the values listed in Table 11-2. In the case of vicinal dibenzoates (entry 1, 2, and 3 in Table 11-2), it was assumed that both benzoate moieties adopt only one conformation in which they are pointing away from each other.

The actual numerical calculation of coordinates of a point dipole is exemplified for the 6α-benzoate group as follows: (i) set plane A containing the benzoate plane and C-5, and C-6 atoms, and place x and y axes in this plane (Figure 11-9); (ii) calculate the local coordinates of atoms C-5,

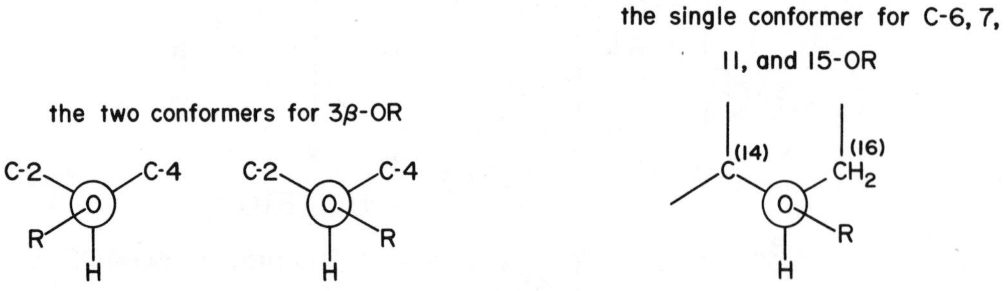

Figure 11-8. *Conformation of the benzoate group OR.* *[Reprinted from reference 3.]*

C-6, ethereal oxygen, and C-7 and the point dipole against x, y, and z axes; then estimate the distance between the point dipole and each atom, respectively; (iii) set the other plane B containing C-3, C-4, C-5, C-6, and C-7 atoms and place X, Y, and Z axes in this plane and obtain the principal coordinates of C-5, C-6, ethereal oxygen, and C-7 against the X, Y, and Z axes; then (iv) this gives four second-order equations of three unknown parameters, X_i, Y_i, and Z_i. The principal coordinates of the point dipole i, X_i, Y_i, and Z_i, are thus calculable.

The absolute value of the electric transition length $|\vec{r}| = |\vec{\mu}|/e$ was estimated from the UV spectra of the corresponding p-dimethylaminobenzoate. The experimental dipole strength D which is theoretically expressed as $D = e^2|\vec{r}|^2$ is related to the molar extinction coefficient ε as follows.

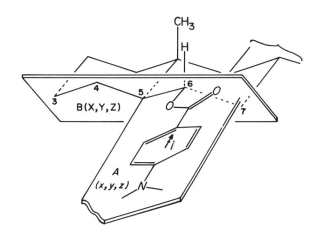

Figure 11-9. *Estimations of the principal coordinates of a point dipole i, X_i, Y_i, Z_i. [Reprinted from reference 3.]*

$$D = 0.9184 \times 10^{-38} \int_0^\infty \epsilon(\sigma)/\sigma \, d\sigma \qquad \text{(cgs unit)} \qquad (11.35)$$

The numerical integration of the actual UV spectrum curve gave $|\vec{r}| = 1.156$ Å.

The interaction energy V_{ij}, which varied from 353.8 cm^{-1} for entry 1 in Table 11-2 to 44.2 cm^{-1} for entry 10, was computed by using the point dipole approximation. The rotational strengths were also computed by equations 10.88 and 10.92; for example, for entry 1, $R^\beta = -0.457 \times 10^{-37}$ cgs unit.

The component CD curve is expressed by

$$\Delta\epsilon(\sigma)^k = \Delta\epsilon_{max}^k \, f(\sigma + \sigma_0 - \sigma_k) \qquad (11.36)$$

where $\Delta\epsilon_{max}^k$ is the maximum value of the kth Cotton effect, and $f(\sigma)$ is the function describing the shape of a component CD Cotton effect curve and is adopted from the observed UV spectra. Then, the rotational strength expressed by equation 10.1 becomes

$$R^k = 2.296 \times 10^{-39} \Delta\epsilon_{max}^k \int_0^\infty f(\sigma)/\sigma \, d\sigma \qquad (11.37)$$

where $f(\sigma + \sigma_0 - \sigma_k)$ is approximated to be $f(\sigma)$ because of the small value of $\sigma_0 - \sigma_k$. From equations 11.36 and 11.37, the following expression for a component CD Cotton effect curve is derived.

$$\Delta\epsilon(\sigma)^k = \frac{R^k}{2.296 \times 10^{-39} \int_0^\infty f(\sigma)/\sigma \, d\sigma} f(\sigma + \sigma_0 - \sigma_k) \qquad (11.38)$$

The actual CD curves frequently shows a conspicuous imbalance in the shape of two Cotton effects; i.e., the first Cotton effect is sharper and stronger than the second Cotton effect. This phenomenon is unexpected if one considers the equality of rotational strengths of the two excited states, α and β: $R^{\alpha} = (1/2)\pi\sigma_0 \vec{R}_{ij} \cdot (\vec{\mu}_{i0a} \times \vec{\mu}_{j0a})$ and $R^{\beta} = -(1/2)\pi\sigma_0 \vec{R}_{ij} \cdot (\vec{\mu}_{i0a} \times \vec{\mu}_{j0a})$. The

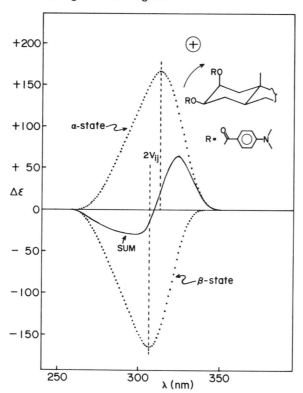

Figure 11-10. Summation of two asymmetric component CD curves gives an asymmetric resultant curve; the example is for 5α-cholestane-2β,3β-diol bis(p-dimethylaminobenzoate). The two component CD curves are derived from experimental UV curves, and hence the summation curve is not smooth. [Adapted from reference 3.]

imbalance between the two apparent Cotton effects cannot be satisfactorily explained by assuming a Gaussian distribution. We therefore adopted the actual shape of the UV band for computing the shape of component CD Cotton effects.

This choice is reasonable because, in the case of coupled Cotton effects due to the exciton chirality mechanism, the magnetic transition moment, which is one of the factors responsible for the rotational strength, originates from the coupled electric moment; i.e., UV excitation. The UV transition naturally reflects the other factor, the electric transition moment, as well.

In general, the experimental UV band plotted against wavelength is asymmetric; namely, it is steeper on the longer wavelength side and broader on the shorter wavelength side. This tendency is further emphasized when the UV spectrum is plotted against the wavenumber. The summation of two such curves of opposite signs separated by the Davydov split $2V_{ij}$ results in narrower first and broader second Cotton effects (Figure 11-10).

The dissymmetrical shape of absorption spectra originates from the Franck-Condon effect in electronic excitation.[4] As illustrated in Figure 11-11 (A), if the equilibrium configuration of excited state is similar to that of the ground state, the 0-0 vibrational band is the most intense because of efficient overlap between vibrational wavefunctions, and intensities for the 0-1, 0-2, 0-3, ..., transitions fall off rapidly. The shape of the absorption curve is extremely dissymmetrical; i.e., steep on longer wavelength side and flat on the shorter wavelength side. The present case is exemplified by the ^1L_a transitions of anthracene (see Figure 1-10) and naphthacene.

(A) with similar equilibrium configuration

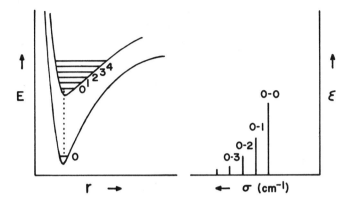

(B) with slightly larger equilibrium configuration

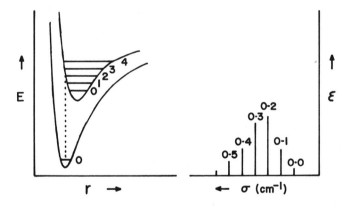

Figure 11-11. *Franck-Condon potential curves of the ground and lowest excited states and absorption spectra: (A) equilibrium configurations of the excited and ground states are similar to each other; (B) the equilibrium configuration in the excited state is slightly larger than that of the ground state. [Adapted from reference 4.]*

On the other hand, if the equilibrium configuration of excited state is slightly larger than that of the ground state, the 0-0 band is no longer the strongest; intensity of each band is changed as illustrated in (B) of Figure 11-11, where the 0-2 band is the most probable transition. In this case, the shape of the whole curve is still dissymmetrical as depicted.   The shape of UV absorption curve is thus generally dissymmetric due to the Franck-Condon principle.

In the case of dimers, the composite CD curve is expressed by summation of the two component Cotton effects α and β, $\Delta\varepsilon(\sigma) = \Delta\varepsilon(\sigma)^{\alpha} + \Delta\varepsilon(\sigma)^{\beta}$, and these two Cotton effects are separated by the so-called Davydov split $2V_{ij}$. Figure 11-10 shows the calculated CD curve of 5α-cholestane-$2\beta,3\beta$-diol bis(p-dimethylaminobenzoate) (entry 1).   The position, amplitude, and shape of the calculated split Cotton effects are in excellent agreement with the observed (Table 11-2).

On the basis of equation 10.106, the apparent rotational strength of first and second Cotton effects of entry 1 can be calculated to be R_{first}= +0.821×10^{-38} and R_{second}= −0.821×10^{-38} cgs unit.   Therefore, although the shapes of two apparent Cotton effects are uneven, the criterion of equality of two rotational strengths — i.e., $|R_{first}| = |R_{second}|$ — is strictly satisfied. This criterion is met satisfactorily for the observed Cotton effects as well: $R_{first} = +0.756×10^{-38}$, $R_{second} = −0.696×10^{-38}$ cgs unit.   Thus, the imbalance in the shape and amplitude of two apparent Cotton effects originates from the asymmetrical distribution of vibrational factors in the corresponding electronic excitation.

Table 11-2 summarizes the calculated and observed results for the ten bis (p-dimethylaminobenzoates).   The calculated values are in good agreement with the observed values, including the unevenness of the first and second Cotton

Table 11-2. Calculated and Observed CD Cotton Effects of Bis(p-dimethylaminobenzoates) of Various Steroid Glycols and Interchromophoric Distance R_{ij}. [Reprinted from reference 3].

Entry	Compound[a]	$\Delta\varepsilon(\lambda,\text{nm})$ (calcd) Conformer I	Conformer II	$\Delta\varepsilon_{calcd}$, nm [b]	A_{calcd} [c]	$\Delta\varepsilon_{obsd}$, nm	A_{obsd} [c]	Solvent [d]	R_{ij}, Å
1	5α-Cholestane-2β,3β-	+64.0(323) −28.5(296)		+64.0(323) −28.5(296)	+92.5	+61.7(320) −33.2(295)	+94.8	E	8.1
2	5α-Cholestane-2α,3α-	+64.0(323) −28.5(296)		+64.0(323) −28.5(296)	+92.5	+61.3(321) −27.1(295)	+88.4	E	8.1
3	Cholest-5-ene-3β,4β-	−64.0(323) +28.5(296)		−64.0(323) +28.5(296)	−92.5	−57.8(321) +35.6(295)	−93.4	E	8.1
4	5α-Cholestane-3β,6β-	−40.0(324) +18.8(295)	−34.4(323) +15.7(294)	−37.2(323) +17.3(295)	−54.5	−37.6(320) +19.2(295)	−56.8	10%D/E	9.9
5	5α-Cholestane-3β,6α-	+40.0(323) −18.3(296)	+48.5(323) −22.0(295)	+44.3(323) −20.2(295)	+64.5	+59.2(319) −30.2(294)	+89.3	20%D/E	9.5
6	5α-Cholestane-3β,7α-	+31.0(323) −14.0(295)	+28.0(323) −13.0(295)	+29.5(323) −13.5(295)	+43.0	+28.5(320) −11.3(295)	+39.8	E	10.6
7	5α-Cholestane-3β,7β-	−5.2(323) +2.3(295)	0.0 0.0	−2.6(323) +1.2(295)	−3.8	−2.8(321) +4.3(300)	−7.1	E	12.4
8	5α-Androstane-3β,11β-	+11.0(323) −5.0(296)	+19.0(323) −8.6(295)	+15.0(323) −6.8(295)	+21.8	+18.8(320) −8.7(294)	+27.5	10%D/E	11.4
9	5α-Pregnane-3β,11α-	−18.5(324) +8.5(295)	−25.2(323) +11.8(295)	−21.9(323) +10.2(295)	−32.1	−35.0(320) +17.7(295)	−52.7	E	10.3
10	D-Homo-5α-androstane-3β,15β-	−12.0(323) +5.5(295)	−15.7(324) +7.4(295)	−13.9(323) +6.5(295)	−20.4	−20.4(319) +6.0(291)	−26.4	E	12.8

a Positions of p-dimethylaminobenzoyloxy groups follow compound name. b Except of entries 1, 2, and 3, $\Delta\varepsilon_{calcd}$ is the average value of two rotational conformers I and II around the 3β-C-O bond (Figure 11-8). c A = $\Delta\varepsilon_1 - \Delta\varepsilon_2$. d E, ethanol; D, dioxane.

effects.   As mentioned earlier, it has been assumed that the benzoate groups in vicinal dibenzoates (entry 1, 2, and 3) adopt only one conformation and hence there is only one theoretical curve.   For other benzoates, there are two sets of theoretical data (Table 11-2).

It is seen that the position of the two split Cotton effects is fixed around 323 nm (calcd) and 320 nm (obsd) for the first extrema and around 295 nm (calcd) and 295 nm (obsd) for the second extrema.   This effect is easily understood by considering equation 10.119; namely, term A, which determined the shape of curves, contains only parameters $\Delta\sigma$ and σ_0 characteristic   of the UV spectrum of the component chromophore, and therefore locations of two

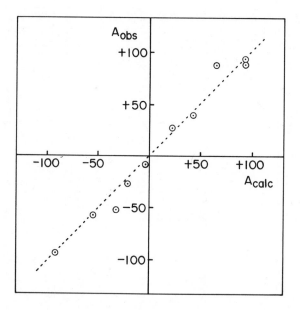

Figure 11-12. *Comparison of calculated and observed A values of bis(p-dimethylaminobenzoates) listed in Table 11-2: (••••••) expected.* *[Adapted from reference 3.]*

apparent Cotton effects are independent of the mutual configuration of the two chromophores.

In the case of 1,8-dibenzoate (entry 10), it is seen that in spite of an interchromophoric distance of 12.8 Å, (distance between the two point dipoles of the 3β- and 15β-benzoates), the amplitudes of A_{calcd} = -20.4 and A_{obsd} = -26.4 are large enough, and therefore the exiton chirality method should be applicable to even more distant binary systems; as mentioned above, the amplitude is inversely proportional to the square of the interchromophoric distance.

Figure 11-12 shows that the observed A values fall within reasonable distances from the ideal linear relationship expected from a plot of A_{obsd} vs. A_{calcd}.

It is notable that the present theoretical treatment based on molecular geometry and experimental UV data reproduces the coupled Cotton effects of binary systems. Although the present data are only for dibenzoate systems, it is clear that a similar treatment should hold for interactions between other chromophores, enone-benzoate, etc., including relatively remote groups.

11-4.   Circular Dichroic Power Due to Chiral Exciton Coupling between Two Polyacene Chromophores[5]

In this section, we discuss the numerical calculation of the CD spectra of (6R,15R)-(+)-6,15-dihydro-6,15-ethanonaphtho[2,3-c]pentaphene (11) and (7R, 14R)-(+)-7,14-dihydro-7,14-ethanodibenz[a,h]anthracene (12), binary systems composed of two anthracene and naphthalene chromophores, respectively.

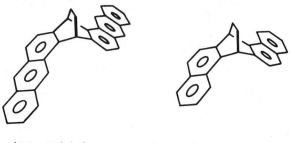

(6R,15R)-(+)-11   **(7R,14R)-(+)-12**

For the theoretical calculation of chiral exciton coupling in compound (+)-11, anthracene was taken as the isolated and component chromophore. From the UV spectrum of anthracene (λ_{max} 252 nm, σ_0= 39 682.5 cm^{-1}, ε 204,000 in EtOH), the transition length r was estimated from equations 10.6 and 10.7, which gave r= 1.915 Å.

In the original mode of coupling, since the two ^1B_b transition moments are antiparallel and are centrally located in the anthracene moieties, the electric interaction energy estimated by the point dipole approximation method is definitely positive. Actually, the value of V_{ij} calculated by equation 10.84 is +1130.9 cm^{-1}. This value is so large that the present exciton coupling belongs to the category of strong coupling. Since V_{ij} is positive, equations in Table 10-2 indicate that the α state is located at longer wavelengths than the β state.

As illustrated in Figure 11-14, the two ^1B_b transition moments are twisted in a clockwise sense. Therefore, the triple product $\vec{R}_{ij} \cdot (\vec{\mu}_{i0a} \times \vec{\mu}_{j0a})$ is definitely positive; namely the rotational strength of the α state is positive, while that of the β state is necessarily negative. The numerical calculation of the theoretical rotational strength was performed by using the Cartesian coordinates of compound (+)-11 and the UV spectral data; this led to R^{α}= +10.0×10^{-38} and R^{β}= -10.0×10^{-38} cgs unit.

As previously discussed, the exciton chirality governing the sign and amplitude of split Cotton effects is defined by the quadruple product $\vec{R}_{ij} \cdot (\vec{\mu}_{i0a} \times \vec{\mu}_{j0a}) V_{ij}$. In the present case, since the quadruple product is positive, the first and second Cotton effects are expected to be positive and

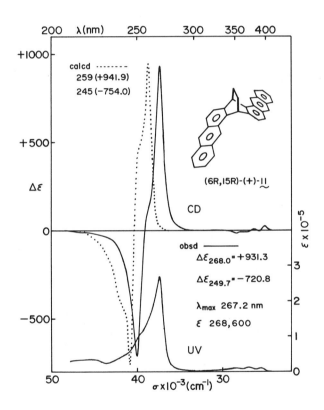

Figure 11-13. *Calculated and observed CD and UV spectra due to the* ^1B_b *transition of hydrocarbon (6R,15R)-(+)-11:* (——), *observed CD;* (••••••), *calculated CD;* (——), *observed UV. [Adapted from reference 5.]*

negative, respectively. The CD spectrum curve of compound (+)-11 was numerically calculated by using equation 11.38 and the shape of the actual UV spectrum of anthracene (Figure 1-10); the calculated CD Cotton effects (λ_{ext} 259 nm, $\Delta\varepsilon$ +941.9 and λ_{ext} 245 nm, $\Delta\varepsilon$ -754.0) are in excellent agreement with the experimental values (λ_{ext} 268 nm, $\Delta\varepsilon$ +931.3 and λ_{ext} 249.7 nm, $\Delta\varepsilon$ -720.8) (Figure 11-13). Namely, not only the sign but also the position, amplitude,

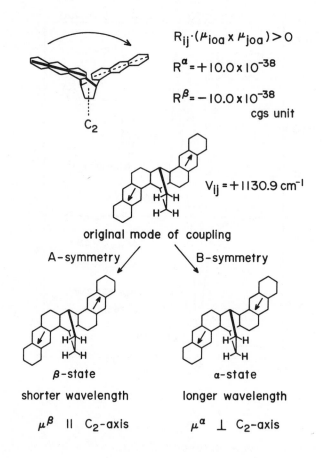

$$R_{ij}\cdot(\mu_{ioa}\times\mu_{joa})>0$$

$$R^{\alpha}=+10.0\times10^{-38}$$

$$R^{\beta}=-10.0\times10^{-38}$$
$$\text{cgs unit}$$

C_2

$$V_{ij}=+1130.9\ cm^{-1}$$

original mode of coupling

A-symmetry   B-symmetry

β-state α-state

shorter wavelength   longer wavelength

μ^{β} ∥ C_2-axis μ^{α} ⊥ C_2-axis

Figure 11-14.   Rotational strength and coupling modes of two electric transition moments for the ^1B_b transition of (+)-11. [Reprinted from reference 5.]

Table 11-3. Spectral and Geometric Parameters and CD Cotton Effects Due to ^1B_b Transition of Compounds (+)-11 and (+)-12. [Reprinted from reference 5].

	(+)-11	(+)-12
Component chromophore	Anthracene	1,2-Dimethyl-naphthalene
σ_0, cm^{-1}	39 682.5	43 956.0
Transition length r, Å	1.915	1.650
R_{ij}, Å	7.689	5.808
V_{ij}, cm^{-1}	+1130.9	+1060.7
Rotational strength$\times 10^{38}$, cgs unit		
Theoretical	R^α +10.0	R^α +8.23
	R^β -10.0	R^β -8.23
Theoretical apparent	R_{1st} +7.73	R_{1st} +4.41
	R_{2nd} -7.80	R_{2nd} -4.43
Observed apparent	R_{1st} +7.43	R_{1st} +3.50
	R_{2nd} -7.53	R_{2nd} -1.70
CD Cotton effects		
Calcd	259 nm, $\Delta\varepsilon$ +941.9	233 nm, $\Delta\varepsilon$ +470.9
	245 nm, $\Delta\varepsilon$ -754.0	221 nm, $\Delta\varepsilon$ -252.6
Obsd	268 nm, $\Delta\varepsilon$ +931.3	237 nm, $\Delta\varepsilon$ +326.5
	249.7nm,$\Delta\varepsilon$ -720.8	224 nm, $\Delta\varepsilon$ -180.5

and shape of calculated Cotton effects are in agreement with experiment.   The
present treatment thus establishes the (6R,15R) absolute stereochemistry of
compound (+)-11 in a nonempirical manner.

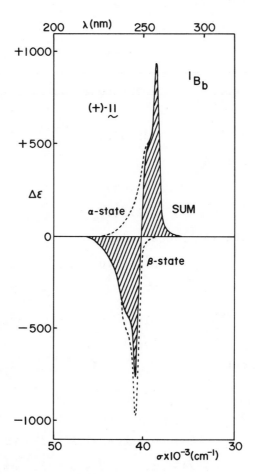

*Figure 11-15.   Summation of two component CD curves (•••••)   taken   from   the
shape of the UV spectrum of anthracene gives the calculated split CD Cotton
effect curve (_____) of compound (+)-11.   Positive and negative shaded areas
correspond to the calculated apparent rotational strengths of first and second
Cotton effects, respectively (see Table 11-3).   [Adapted from reference 5.]*

The following analysis of the observed CD spectra confirms the above assignment. The rotational strengths of the observed apparent Cotton effects were calculated by equation 10.1, giving R(first apparent) $+7.43 \times 10^{-38}$ and R(second apparent) -7.53×10^{-38} cgs unit, where the excitation wavenumber of anthracene, $\sigma_0 = 39,682.5$ cm^{-1}, was employed (Table 11-3 and Figure 11-15).

It is thus seen that the observed Cotton effects exactly obey the sum rule ($\sum R = 0$); this means that the present Cotton effects are based exclusively on the chiral exciton coupling between the two ^1B_b transitions, without participation of other ^1L_a or ^1L_b transitions. Namely, it is valid to employ only the conservative exciton coupling term in this case.

The analysis of the UV spectrum also confirms the exciton coupling mechanism; as shown in Figure 11-14, the phase of two electric transition moments of α state at longer wavelength is in phase, while that of the β state at shorter wavelength is out of phase. Since the angle between the two vectors is 151.0°, the in-phase combination of electric transition moments results in a larger resultant moment than that of the out-of-phase combination (Figure 11-16). The calculated ratio of two dipole strengths is $D^{\alpha}:D^{\beta} = 15:1$.

The theoretical calculation based on the exciton coupling mechanism predicts that the UV of 11 should be more intense at longer wavelengths than at shorter wavelengths. In other words, the UV peak is shifted towards the first Cotton effect at longer wavelength. The observed UV spectrum provides a striking confirmation of this effect; the UV maximum peak is located at 37,425.1 cm^{-1}, which is closer to the first Cotton effect at 37,313.4 cm^{-1} than to the second one at 40,048.0 cm^{-1} (Figure 11-13). From C_2-symmetrical properties, the transition of β state (A symmetry) is polarized along the C_2 axis, while the transition moment of the α state (B symmetry) is perpendicular to this axis (Figures 11-14 and 11-16).

The observed UV spectrum was actually divided, in a 15:1 ratio, into the two component α and β transitions, as shown in Figure 11-16. From this analysis, it is likely that the shoulder around 250 nm is due to the weak transi-

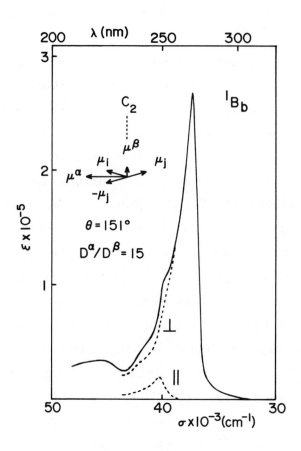

Figure 11-16. *Calculated divided spectra of the ^1B_b transition of hydrocarbon (+)-11. Symbols ⊥ and ‖ indicate the absorption perpendicular and parallel to the C_2 axis, respectively. Since the angle θ between two moments is 151.0°, the ratio μ^α/μ^β is $\sqrt{15}$, therefore D^α/D^β = 15: (——), observed UV; (•••••), divided spectra. [Adapted from reference 5.]*

tion of the β state.   Thus, the uneven shape of the UV spectrum can be rationalized by a simple exciton coupling mechanism.

The CD spectrum of hydrocarbon (+)-<u>12</u> was calculated in a similar manner
using the UV spectral data of 1,2-dimethylnaphthalene (Figure 11-17).
Although the calculated amplitude is larger than the observed, the sign and

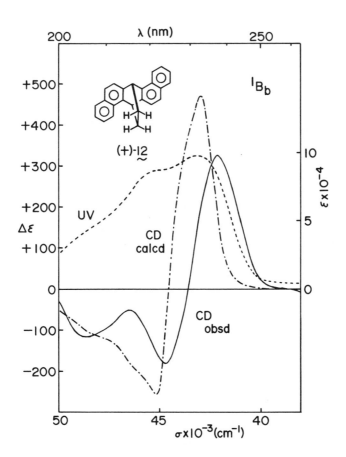

Figure 11-17. *CD and UV spectra of compound (+)-<u>12</u> in the region of ^1B_b tran-*
sition: (___ *), observed CD; (-•-•-), calculated CD; (•••••), observed UV.*
[Adapted from reference 5.]

position of Cotton effects are in good agreement with experiment, which establishes the (7R,14R) absolute configuration of compound (+)-12.

11-5.   Circular Dichroic Power of Chiral Triptycenes[6]

In this section, the circular dichroism of chiral triptycenes, (5R,12R)-
(-)-1,15-diethynyl-5,12-dihydro-5,12[1',2'] benzenonaphthacene (13) and (5S,
12S)-(+)-5,12-dihydro-5,12[1',2'] benzenonaphthacene-1,15-dicarbonitrile (14)
which were qualitatively discussed in Chapter 4 are theoretically calculated
by the coupled oscillator theory.

(5R,I2R)-(−)-I3   (5S,I2S)-(+)-I4

The calculated CD spectra are in excellent agreement with the observed
spectra and corroborate the previous qualitative configurational assignment.

In the case of the triple systems of chiral benzotriptycenes 13 and 14,
the excitation energies of component chromophores, i.e., those of the ^1B_b
transition of naphthalene and the intramolecular charge transfer or ^1L_a
transition of monosubstituted benzene, are almost similar, but not exactly

identical. Therefore, the equation of the rotational strength R^k of kth transition based on the chiral exciton coupling mechanism was modified by adopting the average value of two excitation energies, as follows (see also section 10-9):

$$R^k = (-\pi/2) \sum_{i=1}^{N} \sum_{j>i}^{N} C_{ik}C_{jk}(\sigma_i+\sigma_j)\vec{R}_{ij} \cdot (\vec{\mu}_i \times \vec{\mu}_j) \qquad (11.39)$$

$$(k = 1, 2, 3, \ldots, N)$$

where N is number of chromophores (N=3 in the present case), and σ_i and σ_j are excitation wavenumbers of chromophores i and j, respectively.

The resultant CD spectrum was computed by assuming a Gaussian distribution, because the UV spectra of the two component chromophores — i.e., naphthalene and mono-substituted benzene — differ in shape. The CD curve $\Delta\varepsilon(\sigma)$ is formulated as follows:

$$\Delta\varepsilon(\sigma) = \{(\sigma_i+\sigma_j)/(2\times2.296\times10^{-39}\sqrt{\pi}\Delta\sigma)\} \sum_{k=1}^{N} R^k \exp\{-((\sigma-\sigma_k)/\Delta\sigma)^2\} \qquad (11.40)$$

where standard deviation of the Gaussian distribution was approximated by the average $\Delta\sigma$ value of two component transitions.

In the calculation of chiral exciton coupling in compounds (-)-13 and (+)-14, 2,3-dimethylnaphthalene, ethynylbenzene, and 2-methylbenzonitrile were adopted as the component chromophores.

In the case of the diethynyl compound (-)-13, the exciton wave functions indicated in Figure 11-18 were obtained by solving the secular equation of three dimension. It is noteworthy that the second transition is optically inactive. The second transition is anti-symmetrical against rotation around the C_2-symmetrical axis. Therefore, the wave function Ψ_2 of the second excited state excludes the excitation wave function ϕ_3 of the naphthalene

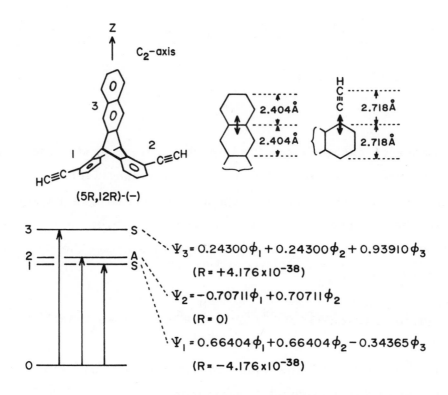

$$\Psi_3 = 0.24300\phi_1 + 0.24300\phi_2 + 0.93910\phi_3$$
$$(R = +4.176 \times 10^{-38})$$

$$\Psi_2 = -0.70711\phi_1 + 0.70711\phi_2$$
$$(R = 0)$$

$$\Psi_1 = 0.66404\phi_1 + 0.66404\phi_2 - 0.34365\phi_3$$
$$(R = -4.176 \times 10^{-38})$$

Figure 11-18. Excitation diagram, wave functions, rotational strengths, and location of excitons of (5R,12R)-(-)-diethynylbenzotriptycene 13. The wave functions Ψ_1, Ψ_2, and Ψ_3 are symmetrical (S), anti-symmetrical (A), and symmetrical (S) around the C_2-symmetrical axis, respectively. [Reprinted from reference 6.]

<u>Table 11-4.</u>   UV Spectral Data [a] of the ^1B_b Transition of 2,3-Dimethyl-naphthalene and the Intramolecular Charge Transfer or ^1L_a Transition of Ethynylbenzene and 2-Methylbenzonitrile. [Reprinted from reference 6.]

Compound	2,3-dimethyl-naphthalene	ethynylbenzene	2-methyl-benzonitrile
λ_{max} (nm)	226.0	234.0	228.0
ε_{max}	92500	15000	14200
σ_0 (cm^{-1})	44247.8	42735.0	43859.6
r (Å)	1.6647	0.7716	0.7374
$\Delta\sigma$ (cm^{-1})	1790.2	2530.5	2503.5

[a] Solvent: EtOH

chromophore.   Since the two transition moments of remaining ethynylbenzene chromophores are parallel to each other, the rotational strength R is nil. Namely, the second transition is optically inactive.   The first and third transitions give the conservative rotational strength as shown in Figure 11-18. The same is true in the case of the dinitrile (+)-<u>14</u>.

The CD spectrum of compound (-)-<u>13</u> was numerically calculated by employing equation 11.40 and the parameters listed in Table 11-4.   The calculated values of CD Cotton effects, λ_{ext} 240.0 nm, $\Delta\varepsilon$ -135.5 and 221.0 nm, $\Delta\varepsilon$ +135.3, are in excellent agreement with the experiment, λ_{ext} 245.5 nm, $\Delta\varepsilon$ −138.2 and 215.0 nm, $\Delta\varepsilon$ +113.6 (Table 11-5 and Figure 11-19). Thus, the present theoretical treatment establishes the (5R,12R) absolute stereochemistry of the diethynyl compound (−)-13 in a nonempirical manner.

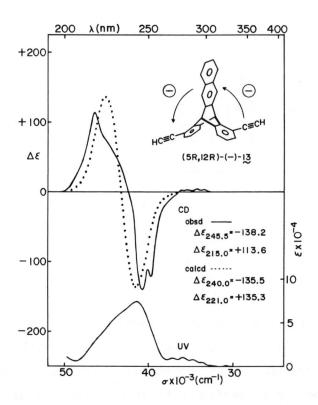

Figure 11-19. _UV and CD spectra of (5R,12R)-(−)-13 in EtOH: UV, λ_{max} 241.0 nm (ε 75 000). Dotted line shows the CD spectrum calculated by application of the exciton chirality method. [Adapted from reference 6.]_

Similarly, as shown in Table 11-5 and Figure 11-20, the excellent agreement between the calculated and observed CD exciton split Cotton effects of (+)-_14_ leads to the unequivocal and nonempirical determination of the (5_S_,12_S_) absolute configuration of the dinitrile (+)-_14_; the result is in line with X-ray data.

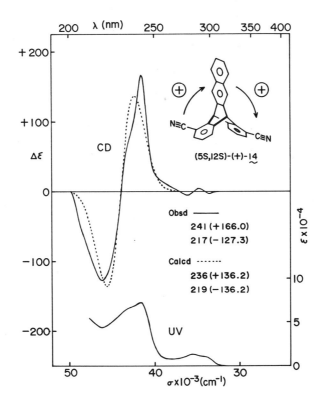

Figure 11-20. UV and CD spectra of (5_S_,12_S_)-(+)-_14_ in EtOH: UV, λ_{max} 241.0 nm (ε 72 400). Dotted line shows the CD spectrum calculated by the CD exciton chirality method. [Adapted from reference 6.]

Table 11-5. The Calculated and Observed CD Cotton Effects of Chiral Benzotriptycenes by the Exciton Chirality Method. [Reprinted from reference 6.]

Compd	Obsd		Calcd	
	UV,λ(nm),(ε)	CD,λ(nm),($\Delta\varepsilon$)	R $\times$ 10^{38},[a]	CD,λ(nm),($\Delta\varepsilon$)
(−)-13	241.0(75,000)	245.5(−138.2)	−4.716	240.0(−135.5)
			0.0	
		215.0(+113.6)		221.0(+135.3)
			+4.716	
(+)-14	241.0(72,400)	241.0(+166.0)	+5.824	236.0(+136.2)
			0.0	
		217.0(−127.3)		219.0(−136.2)
			−5.824	

[a] In cgs unit.

References

1. N. Harada, S. Suzuki, H. Uda, and K. Nakanishi, <u>J. Am. Chem. Soc.</u> <u>93</u>, 5577 (1971).

2. N. Harada and K. Nakanishi, <u>Acc. Chem. Res.</u> <u>5</u>, 257 (1972).

3. N. Harada, S. -M. L. Chen, and K. Nakanishi, <u>J. Am. Chem. Soc.</u> <u>97</u>, 5345 (1975).

4. N. J. Turro, <u>Molecular Photochemistry</u>, (New York: Benjamin, 1965), Chapter 3.

5. N. Harada, Y. Takuma, and H. Uda, <u>J. Am. Chem. Soc.</u> <u>100</u>, 4029 (1978).

6. N. Harada, Y. Tamai, and H. Uda, <u>J. Am. Chem. Soc.</u> <u>102</u>, 506 (1980).

XII. THEORETICAL CALCULATION OF CD SPECTRA BY THE DIPOLE VELOCITY MOLECULAR ORBITAL METHOD

12-1. Concept of the Dipole Velocity Molecular Orbital Method

As discussed in the previous chapters, the exciton chirality method is extensively applicable for determining the absolute configurations of a variety of organic compounds. In some isolated cases, however, observed CD spectra are too complex to be explained by the coupled oscillator mechanism. For example, the complex circular dichroism of chiral tribenzotriptycene, (7S,14S)-(+)-7,14-dihydro-7,14[1',2']naphthalenobenzo[a]naphthacene (1), could not be accounted for by a simple exciton coupling method. In such cases, the self-consistent/configuration interaction/dipole velocity molecular orbital method (SCF-CI-DV MO method) is versatile for theoretical calculation of circular dichroic activity. In this chapter, the SCF-CI-DV MO method and its application to CD and UV spectra are briefly discussed.

As described in the section 10-5, the following equation[1] is significant in the present case. For excitation a→b,

$$\vec{P}_{ab} = -(2\pi imc/e)\,\sigma_{ba}\vec{\mu}_{ab} \qquad (12.1)$$

where

$$\vec{P}_{ab} = \int \phi_a \Sigma \vec{P}_{is} \phi_b d\tau \tag{12.2}$$

$$\vec{\mu}_{ab} = \int \phi_a \Sigma \vec{\mu} \phi_b d\tau \tag{12.3}$$

$$\sigma_{ba} = \nu_{ba}/c = (E_b - E_a)/(hc) \tag{12.4}$$

$\vec{P}_{is}$ is linear momentum operator and is formulated as follows:[2]

$$\vec{P}_{is} = (\frac{h}{2\pi i}) (\vec{i}\frac{\partial}{\partial x} + \vec{j}\frac{\partial}{\partial y} + \vec{k}\frac{\partial}{\partial z}) \tag{12.5}$$

This operator is also expressed as

$$\vec{P}_{is} = (\hbar/i)\vec{\nabla}_{is} \tag{12.6}$$

where $\vec{\nabla}_{is}$ is called the "del operator" and is formulated as

$$\vec{\nabla}_{is} = \vec{i}\frac{\partial}{\partial x} + \vec{j}\frac{\partial}{\partial y} + \vec{k}\frac{\partial}{\partial z} \tag{12.7}$$

$$\hbar = h/(2\pi) \tag{12.8}$$

Substitution of equation 12.6 into equation 12.1 gave

$$\vec{\mu}_{ab} = <a|\vec{\mu}|b> = (\hbar/i)<a|\Sigma\vec{\nabla}_{is}|b>(\frac{-e}{2\pi imc}) (1/\sigma_{ba})$$

$$= <a|\Sigma\vec{\nabla}_{is}|b>\beta_M/(\pi\sigma_{ba}) \tag{12.9}$$

where β_M is the Bohr magneton which is defined as

$$\beta_M = e\hbar/(2mc) \tag{12.10}$$

Therefore,

$$\langle a|\sum e\vec{r}_{is}|b\rangle = \langle a|\sum\vec{V}_{is}|b\rangle\beta_M/(\pi\sigma_{ba}) \tag{12.11}$$

Thus, electric transition moment is expressed by both terms of equation 12.11; the left side term is a representation in "dipole length" form because of length $\vec{r}$, and the right side term is in "dipole velocity" form because of the nature of the $\vec{V}$ operator. When calculating CD spectra, it is recommended that one employ the dipole velocity representation in order to render the rotational strength independent of its origin.[3] From the view point of the dipole-velocity method, the magnetic moment operator is expressed as follows:

$$\vec{M} = (e/2mc)\sum(\vec{r}_{is}\times\vec{p}_{is})$$

$$= (e\hbar/2mci)\sum(\vec{r}_{is}\times\vec{V}_{is}) \tag{12.12}$$

Therefore, the magnetic transition moment is

$$\langle b|\vec{M}|a\rangle = (e\hbar/2mci)\langle b|\sum\vec{r}_{is}\times\vec{V}_{is}|a\rangle$$

$$= (\beta_M/i)\langle b|\sum\vec{r}_{is}\times\vec{V}_{is}|a\rangle \tag{12.13}$$

By combining equations 12.11 and 12.13, rotational strength is formulated as

$$R = \text{Im}\{\langle a|\vec{\mu}|b\rangle\cdot\langle b|\vec{M}|a\rangle\} \tag{12.14}$$

Since magnetic moment operator is Hermitian (see section 10-5),

$$<b|\vec{M}|a> = -<a|\vec{M}|b> \tag{12.15}$$

$$R = Im\{<a|\vec{\mu}|b>\cdot(-1)<a|\vec{M}|b>\}$$

$$= -Im\{<a|\sum\vec{V}_{is}|b>(\beta_M/\pi\sigma_{ba})\cdot(\beta_M/i)<a|\sum\vec{r}_{is}\times\vec{V}_{is}|b>\}$$

$$= <a|\sum\vec{V}_{is}|b>\cdot<a|\sum\vec{r}_{is}\times\vec{V}_{is}|b>\beta_M^{\ 2}/(\pi\sigma_{ba}) \tag{12.16}$$

In common cases, since the molecular orbital corresponding to the excitation a→b is occupied by two electrons, rotational strength is doubled.[4]   Therefore,

$$R = 2<a|\sum\vec{V}_{is}|b>\cdot<a|\sum\vec{r}_{is}\times\vec{V}_{is}|b>\beta_M^{\ 2}/(\pi\sigma_{ba}) \tag{12.17}$$

Similarly, dipole strength is also doubled and expressed as[4]

$$D = 2<a|\sum\vec{V}_{is}|b>^2\beta_M^{\ 2}/(\pi\sigma_{ba})^2 \tag{12.18}$$

Thus, if molecular orbital wave functions of ground a and excited b states are obtainable, rotational and dipole strengths are calculable by employing equations 12.17 and 12.18.

Electric and magnetic transition moment terms,

$$<a|\sum\vec{V}_{is}|b> \quad and \quad <a|\sum\vec{r}_{is}\times\vec{V}_{is}|b>$$

are reduced to atomic orbital levels as follows: the z axis components of electric and magnetic moment terms are formulated as[4]

$$\langle a | \sum \vec{\nabla}_{is} | b \rangle_z = \sum_{bonds} (C_{ra}C_{sb} - C_{sa}C_{rb}) \langle \nabla_{rs} \rangle \cos Z_{rs} \tag{12.19}$$

$$\langle a | \sum \vec{r}_{is} \times \vec{\nabla}_{is} | b \rangle_z = \sum_{bonds} (C_{ra}C_{sb} - C_{sa}C_{rb}) \langle \nabla_{rs} \rangle (X_{rs} \cos Y_{rs} - Y_{rs} \cos X_{rs}) \tag{12.20}$$

$$\cos Z_{rs} = (Z_r - Z_s)/R_{rs} \tag{12.21}$$

$$X_{rs} = (X_r + X_s)/2 \tag{12.22}$$

where $\sum_{bonds}$ denotes summation over bonds r-s, namely, moments between non-neighboring atoms are neglected, C_{ra} is the coefficient of atomic orbital r in the wave function Ψ_a, $\langle \nabla_{rs} \rangle$ is the expectation value of a dipole velocity vector $\vec{\nabla}_{rs}$ which is directed along the bond rs in the direction r→s, X_r and Z_r are the x and z coordinates of atom r, respectively, and R_{rs} is the length of bond r-s. The x and y components of the electric and magnetic transition moments are calculable in a similar manner.

12-2.   Theoretical Calculation of Del Value

The del value $\langle \nabla_{rs} \rangle$, a key parameter of the dipole velocity molecular orbital method, is theoretically calculable[5] on the basis of the Slater orbital, as in the case of overlap integral.[6]   The calculation procedure is briefly accounted for in this section.

For a π-bond depicted in Figure 12-1, del value $\langle \nabla_{ab} \rangle$ is calcualted as follows:

$$\langle \nabla_{ab} \rangle \equiv |\vec{\nabla}_{ab}| = \left| \int \chi_{2px}^{a} \vec{\nabla} \chi_{2px}^{b} \, d\tau \right| \qquad (12.23)$$

where χ_{2px}^{a} and χ_{2px}^{b} are $2p\pi$ atomic orbitals of atoms a and b, respectively, and are approximated by Slater orbitals, and the del operator $\vec{\nabla}$ is defined as

$$\vec{\nabla} = \vec{i}\frac{\partial}{\partial x} + \vec{j}\frac{\partial}{\partial y} + \vec{k}\frac{\partial}{\partial z} \qquad (12.24)$$

The $2p\pi$ Slater orbital is represented by Cartesian coordinates as follows;

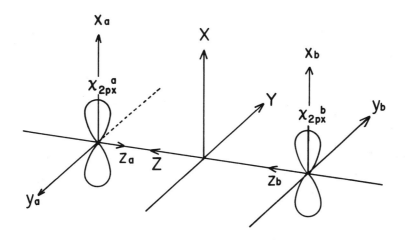

Figure 12-1. *Cartesian coordinate system of two $2p\pi$ orbitals χ_{2px}^{a} and χ_{2px}^{b}.*

$$X_{2px} = \frac{1}{4\sqrt{2\pi}}(\frac{Z^*}{a_0})^{5/2}\exp\{-(\frac{Z^*}{2a_0})\,(x^2+y^2+z^2)^{1/2}\}x \tag{12.25}$$

where Z^* is the effective nuclear charge, and a_0, the Bohr radius is expressed as

$$a_0 = \hbar^2/(me^2) = 0.5292 \; \overset{\circ}{A} \tag{12.26}$$

Operation on atomic orbital X_{2px}^b with del operator $\vec{\nabla}$ is calculated as

$$\vec{\nabla}X_{2px}^b = \frac{1}{4\sqrt{2\pi}}(\frac{Z_b^*}{a_0})^{5/2}(\vec{i}\frac{\partial}{\partial x} +\vec{j}\frac{\partial}{\partial y} +\vec{k}\frac{\partial}{\partial z})\exp\{-(\frac{Z_b^*}{2a_0})\,(x_b^2+y_b^2+z_b^2)^{1/2}\}x_b$$

$$= \frac{1}{4\sqrt{2\pi}}(\frac{Z_b^*}{a_0})^{5/2}(\vec{i}\frac{\partial}{\partial x_b} +\vec{j}\frac{\partial}{\partial y_b} +\vec{k}\frac{\partial}{\partial z_b})\exp\{-(\frac{Z_b^*}{2a_0})\,(x_b^2+y_b^2+z_b^2)^{1/2}\}x_b$$

$$= \frac{1}{4\sqrt{2\pi}}(\frac{Z_b^*}{a_0})^{5/2}$$

$$\times\left[\; \vec{i}\,[\exp\{-(\frac{Z_b^*}{2a_0})\,(x_b^2+y_b^2+z_b^2)^{1/2}\}\,(-\frac{Z_b^*}{2a_0})\,(x_b^2+y_b^2+z_b^2)^{-1/2}\,(x_b)^2\right.$$

$$+\exp\{-(\frac{Z_b^*}{2a_0})\,(x_b^2+y_b^2+z_b^2)^{1/2}\}]$$

$$+\vec{j}\,[\exp\{-(\frac{Z_b^*}{2a_0})\,(x_b^2+y_b^2+z_b^2)^{1/2}\}\,(-\frac{Z_b^*}{2a_0})\,(x_b^2+y_b^2+z_b^2)^{-1/2}\,(x_by_b)]$$

$$+\vec{k}\left[\exp\left\{-\left(\frac{Z_b^*}{2a_0}\right)(x_b^2+y_b^2+z_b^2)^{1/2}\right\}\left(-\frac{Z_b^*}{2a_0}\right)(x_b^2+y_b^2+z_b^2)^{-1/2}(x_b z_b)\right]$$

$$= \vec{i}\left\{\chi_{2px}^b\left(-\frac{Z_b^*}{2a_0}\right)\left(\frac{x_b}{r_b}\right)+\chi_{2px}^b\left(\frac{1}{x_b}\right)\right\} +\vec{j}\left\{\chi_{2px}^b\left(-\frac{Z_b^*}{2a_0}\right)\left(\frac{y_b}{r_b}\right)\right\}$$

$$+\vec{k}\left\{\chi_{2px}^b\left(-\frac{Z_b^*}{2a_0}\right)\left(\frac{z_b}{r_b}\right)\right\} \tag{12.27}$$

where principal coordinates (x, y, and z) and local coordinates of atom b (x_b, y_b, and z_b) are defined as depicted in Figure 12-1.

Substitution of equation 12.27 into equation 12.23 gives

$$\vec{\nabla}_{ab} = \vec{i}\left\{\int\chi_{2px}^a\chi_{2px}^b(-z_b^*/2a_0)(x_b/r_b)d\tau+\int\chi_{2px}^a\chi_{2px}^b(1/x_b)d\tau\right\}$$

$$+\vec{j}\int\chi_{2px}^a\chi_{2px}^b(-z_b^*/2a_0)(y_b/r_b)d\tau$$

$$+\vec{k}\int\chi_{2px}^a\chi_{2px}^b(-z_b^*/2a_0)(z_b/r_b)d\tau \tag{12.28}$$

From the symmetrical nature of χ_{2px}^a, χ_{2px}^b, x_b, and y_b, it is easily under-standable that $\vec{i}$ and $\vec{j}$ vector terms vanish by integration, because products $\chi_{2px}^a\chi_{2px}^b x_b$, $\chi_{2px}^a\chi_{2px}^b/x_b$, and $\chi_{2px}^a\chi_{2px}^b y_b$ have yz, yz, and xz nodal planes, re-spectively. On the other hand, the $\vec{k}$ vector term takes a non-zero value. Therefore,

$$\vec{V}_{ab} = \vec{k} \int \chi_{2px}^{a} \chi_{2px}^{b} (-z_b^{*}/2a_0)(z_b/r_b)\,d\tau \tag{12.29}$$

The integration is calculable as in the case of the calculation of over-lap integrals performed by Mulliken and coworkers.[6]

The polar coordinate system is defined as depicted in Figure 12-2, from which the following relations are derived.

$$x_b = r_b \sin\theta_b \cos\phi_b \tag{12.30}$$

$$y_b = r_b \sin\theta_b \cos\phi_b \tag{12.31}$$

$$z_b = r_b \cos\theta_b \tag{12.32}$$

$$x_a = r_a \sin\theta_a \cos\phi_a \tag{12.33}$$

$$y_a = r_a \sin\theta_a (-\sin\phi_a) \tag{12.34}$$

$$z_a = r_a \cos\theta_a \tag{12.35}$$

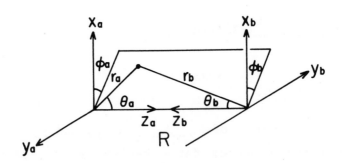

Figure 12-2. *Polar and spheroidal coordinates systems of two atomic orbitals* χ_{2px}^{a} *and* χ_{2px}^{b} *separated by distance R.*

By using the above relations, the Slater functions $\chi_{2px}{}^{a}$ and $\vec{\nabla}\chi_{2px}{}^{b}$ are express-ed as follows;

$$\chi_{2px}{}^{a} = \frac{1}{4\sqrt{2\pi}}(\frac{2\mu_a}{a_0})^{5/2}\exp\{-(\frac{\mu_a r_a}{a_0})\}r_a\sin\theta_a\cos\phi_a \qquad (12.36)$$

$$\vec{\nabla}\chi_{2px}{}^{b} = -\frac{\vec{k}}{8\sqrt{2\pi}}(\frac{2\mu_b}{a_0})^{7/2}\exp\{-(\frac{\mu_b r_b}{a_0})\}r_b\cos\theta_b\sin\theta_b\cos\phi_b \qquad (12.37)$$

where

$$\mu_a = Z_a{}^*/2, \qquad\qquad \mu_b = Z_b{}^*/2 \qquad (12.38)$$

Now, the polar coordinates are further transformed to spheroidal coordi-nates ξ, η, and ϕ, as follows:[6]

$$\xi = (r_a+r_b)/R, \qquad 1 \leq \xi < \infty \qquad (12.39)$$

$$\eta = (r_a-r_b)/R, \qquad -1 \leq \eta \leq 1 \qquad (12.40)$$

$$\phi = \phi_a = \phi_b, \qquad 0 \leq \phi \leq 2\pi \qquad (12.41)$$

where R is the distance between atoms a and b. From the above relations,

$$r_a = (\xi + \eta) R/2 \tag{12.42}$$

$$r_b = (\xi - \eta) R/2 \tag{12.43}$$

$$\cos\theta_a = (1 + \xi\eta)/(\xi + \eta) \tag{12.44}$$

$$\cos\theta_b = (1 - \xi\eta)/(\xi - \eta) \tag{12.45}$$

$$\sin\theta_a = [(\xi^2 - 1)(1 - \eta^2)]^{1/2}/(\xi + \eta) \tag{12.46}$$

$$\sin\theta_b = [(\xi^2 - 1)(1 - \eta^2)]^{1/2}/(\xi - \eta) \tag{12.47}$$

The volume element $d\tau$ is expressed by

$$d\tau = \frac{R^3}{8}(\xi^2 - \eta^2)d\xi d\eta d\phi \tag{12.48}$$

Furthermore, the following new parameters p and t are introduced;

$$p = \frac{1}{2}(\mu_a + \mu_b)\frac{R}{a_0} \tag{12.49}$$

$$t = (\mu_a - \mu_b)/(\mu_a + \mu_b) \tag{12.50}$$

From equations 12.49 and 12.50,

$$\mu_a = \frac{a_0 P}{R}(1+t) \tag{12.51}$$

$$\mu_b = \frac{a_0 P}{R}(1-t) \tag{12.52}$$

Therefore, the del value is expressed as

$$\vec{\nabla}_{ab} = \int \chi_{2px}^{a} \vec{\nabla} \chi_{2px}^{b} d\tau$$

$$= -\vec{k}\int \frac{1}{64\pi}(\frac{2\mu_a}{a_0})^{5/2}(\frac{2\mu_b}{a_0})^{7/2} \exp\{-(\frac{\mu_a r_a + \mu_b r_b}{a_0})\}$$

$$r_a r_b \sin\theta_a \cos\theta_b \sin\theta_b \cos\phi_a \cos\phi_b \, d\tau$$

$$= -\vec{k}\int\int \frac{1}{64\pi}\{\frac{2a_0 P}{a_0 R}(1+t)\}^{5/2}\{\frac{2a_0 P}{a_0 R}(1-t)\}^{7/2}$$

$$\exp[-\{\frac{a_0 P}{R}(1+t)\frac{(\xi+\eta)R}{2a_0} + \frac{a_0 P}{R}(1-t)\frac{(\xi-\eta)R}{2a_0}\}]$$

$$\frac{R}{2}(\xi+\eta)\frac{R}{2}(\xi-\eta)\frac{\{(\xi^2-1)(1-\eta^2)\}^{1/2}}{\xi+\eta}\frac{(1-\xi\eta)}{\xi-\eta}$$

$$\frac{\{(\xi^2-1)(1-\eta^2)\}^{1/2}}{\xi-\eta} \cos^2\phi \frac{R^3}{8} (\xi^2-\eta^2) d\xi d\eta d\phi$$

$$= -\vec{k}(32\pi R)^{-1} p^6 (1+t)^{5/2} (1-t)^{7/2} \iint e^{-p(\xi+t\eta)} (\xi^2-1)$$

$$(1-\eta^2)(1-\xi\eta)(\xi+\eta)d\xi d\eta \int_0^{2\pi} \cos^2\phi d\phi \qquad (12.53)$$

Since

$$\int_0^{2\pi} \cos^2\phi d\phi = \pi \qquad (12.54)$$

$$\vec{V}_{ab} = -\vec{k}(32R)^{-1} p^6 (1+t)^{5/2} (1-t)^{7/2}$$

$$\iint e^{-p(\xi+t\eta)} (\xi^2-1)(1-\eta^2)(1-\xi\eta)(\xi+\eta)d\xi d\eta$$

$$= -\vec{k}(32R)^{-1} p^6 (1+t)^{5/2} (1-t)^{7/2}$$

$$\iint e^{-p(\xi+t\eta)} (\xi^4\eta^3-\xi^4\eta+\xi^3\eta^4-2\xi^3\eta^2+\xi^3-2\xi^2\eta^3+2\xi^2\eta$$

$$-\xi\eta^4+2\xi\eta^2-\xi+\eta^3-\eta)\,d\xi d\eta$$

$$= -\vec{k}(32R)^{-1}p^6(1+t)^{5/2}(1-t)^{7/2}[A_4B_3-A_4B_1+A_3B_4-2A_3B_2$$

$$+A_3B_0-2A_2B_3+2A_2B_1-A_1B_4+2A_1B_2-A_1B_0+A_0B_3-A_0B_1]$$

$$= -\vec{k}(32R)^{-1}p^6(1+t)^{5/2}(1-t)^{7/2}$$

$$[(A_4-2A_2+A_0)(B_3-B_1)+(A_3-A_1)(B_4-2B_2+B_0)] \qquad (12.55)$$

where integrals $A_k(p)$ and $B_k(pt)$ are defined and calculable as [6]

$$A_k(p) \equiv \int_1^\infty \xi^k e^{-p\xi}d\xi = e^{-p}\sum_{\mu=1}^{k+1}[k!/p^\mu(k-\mu+1)!] \qquad (12.56)$$

$$B_k(pt) \equiv \int_{-1}^1 \eta^k e^{-pt\eta}d\eta = -e^{-pt}\sum_{\mu=1}^{k+1}[k!/(pt)^\mu(k-\mu+1)!]$$

$$-e^{pt}\sum_{\mu=1}^{k+1}[(-1)^{k-\mu}k!/(pt)^\mu(k-\mu+1)!] \qquad (12.57)$$

Table 12-1.   Slater μ-Values for Valence Shell ns and np Atomic Orbitals. [Adapted from reference 6.]

H	1.00	C^-	1.45	C	1.625
C^+	1.80	N	1.95	N^+	2.125
O^-	2.10	O	2.275	O^+	2.45
F	2.60	Si	1.383	P	1.60
S	1.817	Cl	2.033	Br	2.054
I	1.90				

where, if t=0,

$$B_k(0) = 2/(k+1) \qquad \text{for k even,} \qquad\qquad (12.58)$$

$$= 0 \qquad \text{for k odd} \qquad\qquad (12.59)$$

The bond dipole velocity values, $\langle \nabla_{ab} \rangle$ are thus theoretically calculable.

   The Slater μ-values[6] for valence shell ns and np atomic orbitals are listed in Table 12-1.

12-3. Numerical Calculation of Bond Dipole Velocity Value

By use of equation 12.55, we can now compute the del value of aromatic C–C π bond. Since both atoms are homogeneous,

$$t = 0 \tag{12.60}$$

Therefore,

$$B_3(0) = B_1(0) = 0 \tag{12.61}$$

$$B_4(0) = 2/5 \tag{12.62}$$

$$B_2(0) = 2/3 \tag{12.63}$$

$$B_0(0) = 2 \tag{12.64}$$

$\vec{V}_{ab}$ is reduced to

$$\vec{V}_{ab} = -\vec{k}(32R)^{-1}p^6(A_3-A_1)(16/15)$$

$$= -\vec{k}(30R)^{-1}p^6(A_3-A_1) \tag{12.65}$$

$$A_3 = e^{-p}\left(\frac{3!}{p^1 3!} + \frac{3!}{p^2 2!} + \frac{3!}{p^3 1!} + \frac{3!}{p^4 0!}\right)$$

$$= e^{-p}(p^{-1}+3p^{-2}+6p^{-3}+6p^{-4}) \tag{12.66}$$

$$A_1 = e^{-p}\left(\frac{1!}{p^1 1!} + \frac{1!}{p^2 0!}\right)$$

$$= e^{-p}(p^{-1}+p^{-2}) \tag{12.67}$$

Therefore,

$$\vec{V}_{ab} = -\vec{k}(30R)^{-1}p^6 e^{-p}(2p^{-2}+6p^{-3}+6p^{-4})$$

$$= -\vec{k}(15R)^{-1}e^{-p}(p^4+3p^3+3p^2) \tag{12.68}$$

Now, the values of parameters are

$$R = 1.388\times10^{-8} \text{ cm} \tag{12.69}$$

$$\mu = 1.625 \tag{12.70}$$

$$a_H = 0.529\times10^{-8} \text{ cm} \tag{12.71}$$

Therefore,

$$p = 4.2637 \tag{12.72}$$

$$\vec{V}_{ab} = -\vec{k}(15\times1.388\times10^{-8})^{-1}e^{-p}(p^4+3p^3+3p^2)$$

$$= (-\vec{k})\,4.173 \times 10^7 \ cm^{-1} \qquad\qquad (12.73)$$

Thus, the theoretical del value of the aromatic C–C π bond is

$$<\nabla_{C-C}>(arom., \ theor.) = 4.173 \times 10^7 \ cm^{-1} \qquad\qquad (12.74)$$

The vector $\vec{\nabla}_{ab}$ is directed along the C–C bond in the direction of a→b because of the minus sign (for definition of coordinates see Figure 12-1). Similarly, the del values of other bonds are calculable.

12-4.   Empirical Del Value Estimated from Observed UV Spectra[7]

Because of the semiempirical nature of the Pariser-Parr-Pople molecular orbital method, an empirical del value estimated from UV spectra is desirable. In fact, the empirical del value of aromatic C–C π bond estimated from the excitation wavenumber and the dipole strength of the β band of benzene was reported to be $6.44 \times 10^7 \ cm^{-1}$. However, the value is too large to explain the absorption intensity of the observed UV spectra of several common aromatic compounds. Therefore, instead of the β band of benzene, the ^1B_b transition of anthracene was chosen for estimating the empirical del value; this was because the ^1B_b transition of anthracene around 250 nm is very intense (ε 2×10^5) and is isolated from other transitions (Figure 12-3). These factors lead to a facile and reliable determination of the empirical del value.

Integration of the peak area of the ^1B_b transition employing equation 10.6 gives an observed dipole strength of the following value.

$$D = 8.789 \times 10^{-35} \text{ cgs unit} \qquad (12.75)$$

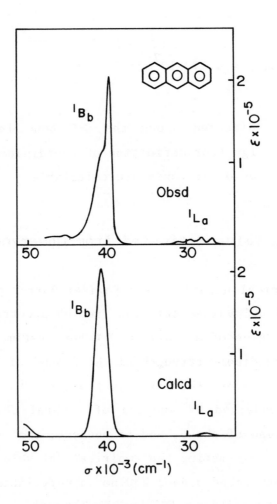

Figure 12-3. Observed and calculated UV spectra of anthracene in ethanol.

Substitution of this value into equation 12.18 and molecular orbital calculation gives

$$\langle \nabla_{C-C} \rangle (\text{arom., empir.}) = 4.701 \times 10^7 \text{ cm}^{-1} \tag{12.76}$$

12-5.   SCF-CI-DV Molecular Orbital Calculation of CD and UV Spectra of Chiral Triptycenes and Related Compounds[7]

The SCF-CI-DV molecular orbital method described in the previous sections is applicable to chiral triptycenes 1, 2, and 3, and related compounds 4 and 5.[8]

In the SCF-CI molecular orbital calculation, configuration interaction between singly excited states of lower energy (maximum 50) was included. The component CD and UV curves were approximated by the Gaussian distribution as in the case of the exciton coupling calculation, and the standard deviation $\Delta\sigma$ values were taken from the observed UV spectra of component chromophores.[1] For example, the $\Delta\sigma$ values of the ^1B_b and ^1L_a transitions of alkyl-substituted naphthalenes employed for the calculation of tribenzotriptycene (+)-1 are 1751.4 and 2853.5 cm^{-1}, respectively. In the case of diethynylbenzotriptycene (−)-2 composed of different chromophores, the average $\Delta\sigma$ value of the, substituted naphthalenes and ethynylbenzene was adopted.

In the calculation of aromatic hydrocarbons, the following atomic orbital parameters were employed: W_C= −11.42 eV, $(rr \mid rr)$= 10.84 eV, β_{C-C} (aromatic)= −2.39 eV. The del value of the aromatic C-C bond $\langle \nabla_{C-C} \rangle$ was estimated to be 4.701×10^7 cm^{-1} from the observed UV spectra of the ^1B_b transition of anthracene in ethanol. The electronic repulsion integral $(rr \mid ss)$ between two atoms

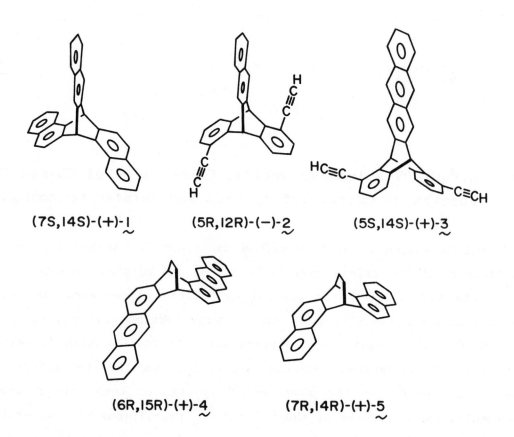

(7S,14S)-(+)-1   (5R,12R)-(−)-2   (5S,14S)-(+)-3

(6R,15R)-(+)-4   (7R,14R)-(+)-5

r and s in one chromophore was estimated by the Nishimoto–Mataga equation. In the case of the diethynyl compound (−)-2, the resonance integral and del values of the acetylene moiety were calculated from the following equations, respectively:

$$\beta = \{s/s_{C-C}(\text{arom.})\}\beta_{C-C}(\text{arom.}) \qquad (12.77)$$

$$\langle V \rangle = \{\langle V \rangle (\text{arom., empir.})/\langle V \rangle (\text{arom., theor.})\}\langle V \rangle (\text{theor.}) \qquad (12.78)$$

where overlap integral S and $\langle V \rangle$ (theor.) were calculable on the basis of the Slater orbitals.

The repulsion integral $(rr \mid ss)$ of two atoms r and s belonging to different chromophores was also estimated by the N-M equation; however, only the distance factor was considered, the angular factor being neglected. The resonance integral value $\beta_{h.c.}$ of the interchromophoric homoconjugation was changed until satisfactory agreement between observed and calculated CD curves was obtained. On the other hand, the $\langle V \rangle$ value of the homoconjugation was completely neglected.

The numerically calculated CD and UV spectra of tribenzotriptycene (+)-<u>1</u> are illustrated in Figure 12-4, where the observed curves are shown in the top. When the interchromophoric resonance is neglected, <u>i.e.</u>, $\beta_{h.c.}$ = 0, the calculated CD curve exhibits two intense Cotton effects of positive first and negative second signs in the ^1B_b transition region.   This situation corresponds to the case of chiral exciton coupling reflecting the combination of two positive and one negative exciton chiralities between the three ^1B_b transition moments.

As the integral $\beta_{h.c.}$ gradually increases, the CD spectrum becomes complex and when $\beta_{h.c.}/\beta_{arom.}$ = 32%, $\beta_{arom.}$ = -2.39 eV, good agreement between the observed and calculated curves was obtained (Table 12-2 and Figure 12-4). Namely, the calculated CD spectrum exhibits four intense Cotton effects of +/-/+/- signs in the region above 34×10^3 cm^{-1}; this is in line with the observed signs, but the calculated amplitudes are smaller than experimental ones. The weak negative Cotton effect around 32×10^3 cm^{-1} is also reproducible by the present calculation (Table 12-2 and Figure 12-4). Moreover, the calculation of the UV curve satisfactorily reproduces the new absorption band of

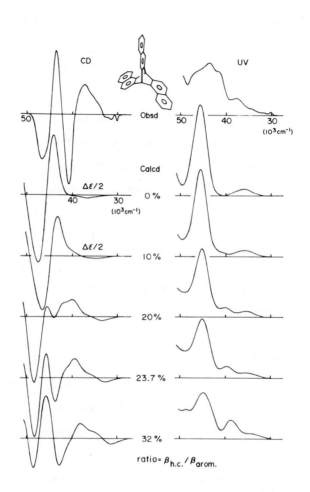

Figure 12-4.   *The CD and UV spectra of (7S,14S)-(+)-tribenzotriptycene 1 obtained by the SCF-CI-DV molecular orbital calculation, in which the value of the interchromophoric homoconjugation resonance integral $\beta_{h.c.}$ was gradually changed. The parameter $\beta_{arom.}$ is the resonance integral of a regular aromatic C-C bond (-2.39 eV). The top curves are the observed CD and UV spectra. The amplitude of the second and third CD curves are illustrated in $\Delta\varepsilon/2$ scale. [Reprinted from reference 7.]*

Table 12-2.   The Observed and Calculated UV and CD Spectra of Chiral Triptycenes and Related Compounds by the SCF–CI–DV Molecular Orbital Method. [Reprinted from reference 7.]

Compd	Obsd		Ratio[a]	Calcd	
	UV,λ(nm),(ε)	CD,λ(nm),($\Delta\varepsilon$)	(%)	UV,λ(nm),(ε)	CD,λ(nm),($\Delta\varepsilon$)
(+)-1		317.0(-24.6)	32		320.5(-29.0)
	264.5(35 800)	267.5(+160.8)		259.0(39 800)	261.8(+78.4)
		244.5(-381.9)			232.5(-142.0)
	229.0(113 800)	229.0(+344.9)		223.2(98 700)	218.3(+226.2)
		213.0(-246.4)			205.7(-157.3)
(-)-2	278.0(10 300)	290.0(+3.5)	20	277.8(2 300)	279.3(+0.6)
	241.0(75 000)	245.5(-138.2)		225.2(92 900)	235.8(-136.8)
		215.0(+113.6)			219.3(+139.0)
(+)-4	371.2(11 200)	397.2(+26.4)	23.7	357.1(11 900)	381.7(+13.5)
		352.8(-14.5)			347.2(-24.2)
	267.2(268 600)	268.0(+931.3)		247.5(299 200)	250.0(+515.2)
		249.7(-720.8)			239.2(-858.4)
(+)-5		304.5(+25.5)	20		295.9(+12.1)
	283.5(11 100)	283.0(-35.6)		279.3(6 700)	261.8(-4.3)
	232.3(98 200)	237.0(+326.5)		220.3(123 500)	223.2(+236.6)
		224.0(-180.5)			211.0(-106.3)

[a] Ratio= $\beta_{h.c.}/\beta_{arom.}$, and $\beta_{arom.}$= -2.39 eV.

medium intensity around 38×10^3 cm^{-1} (Figure 12-4 and Table 12-2).   Thus, the interchromophoric homoconjugation effect plays an important role in the circular dichroism of tribenzotriptycene (+)-1, and the SCF-CI-DV molecular orbital calculation establishes the (7S,14S) absolute stereochemistry.

In a similar way, the CD and UV spectra of diethynylbenzotriptycene (−)-2 were computed.   When $\beta_{h.c.}/\beta_{arom.}$ = 20%, a good agreement between the observed and calculated curves was obtained as shown in Figure 12-5; this establishes the (5R,12R) configuration of (−)-2 in a quantitative manner (Table 12-2).   In this case, it is noteworthy that the basic pattern of the CD Cotton effects around 42×10^3 cm^{-1} — i.e., negative first and positive second signs, — remains unchanged irrespective of the variation in the $\beta_{h.c.}$ value.   This result demonstrates that the circular dichroism of (−)-2 is relatively insensitive to variation of the interchromophoric homoconjugation resonance integral.   In other words, the interchromophoric homoconjugation effect can be neglected, and this in turn corroborates the applicability of the exciton chirality method to the circular dichroism of (−)-2.   Thus, diethynylbenzotriptycene (−)-2 is ideally suited for demonstrating the applicability of the CD exciton chirality method to triple systems.

The SCF-CI-DV molecular orbital calculation method was also applied to a triptycene system composed of anthracene and two ethynylbenzene moieties, (5S,14S)-(+)-1,17-diethynyl-7,14-dihydro-7,14[1',2']benzenopentacene (3), in the same manner as in the case of compound 2.   The calculated curves are in good agreement with the observed CD and UV spectra, when the ratio $\beta_{h.c.}/\beta_{arom.}$ is 20% (Figure 12-6).[10]

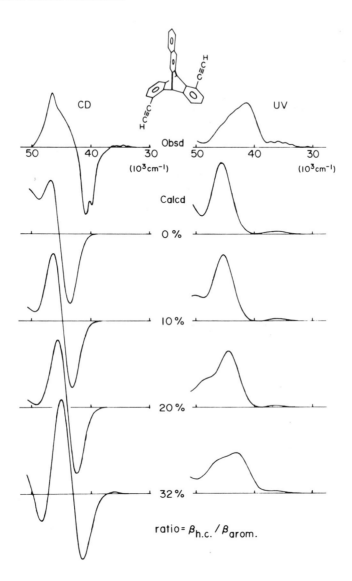

Figure 12-5. The CD and UV spectra of (5R,12R)-(-)-2 obtained by the SCF-CI-DV molecular orbital calculation, in which the value of interchromophoric homoconjugation resonance integral was changed. [Reprinted from reference 7.]

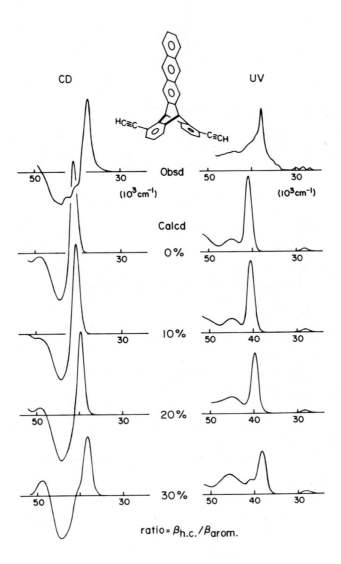

Figure 12-6. *The CD and UV spectra of (5S,14S)-(+)-1,17-diethynyl-7,14-di-hydro-7,14[1',2']benzenopentacene (3) calculated by the SCF-CI-DV molecular orbital method. The top curves are the observed spectra.[10]*

Similarly, the interchromophoric homoconjugation effect has little influence on the exciton circular dichroism of bis-anthracene 4 and bis-naphthalene 5 compounds.   In (+)-4, the pattern of the two Cotton effects of positive first and negative second signs around 40×10^3 cm^{-1}, which are due to the ^1B_b transition of anthracene chromophores, is only slightly affected by variation of the ratio $\beta_{h.c.}/\beta_{arom.}$ (Figure 12-7).   When the ratio is 23.7%, the calculated CD spectrum including the weak positive and negative Cotton effects in the ^1L_a transition region is in good agreement with the observed spectrum (Table 12-2).   Similarly, in the case of (+)-5, a 20% ratio of $\beta_{h.c.}/\beta_{arom.}$ led to a good agreement (Figure 12-8 and Table 12-2).

From the present calculations, the resonance integral $\beta_{h.c.}$ of the interchromophoric homoconjugation in triptycene systems was estimated to be 20–30% of the resonance integral of a regular aromatic C–C bond (Table 12-2).   This value (−0.478 to −0.717 eV) is reasonable in comparison with the theoretical value of −0.567 eV derived from equation 12.77.

The SCF-CI-DV molecular orbital calculations including the interchromophoric homoconjugation effect satisfactorily reproduce the CD spectra of chiral triptycenes and related compounds; i.e., the calculations establish the absolute stereochemistries quantitatively.

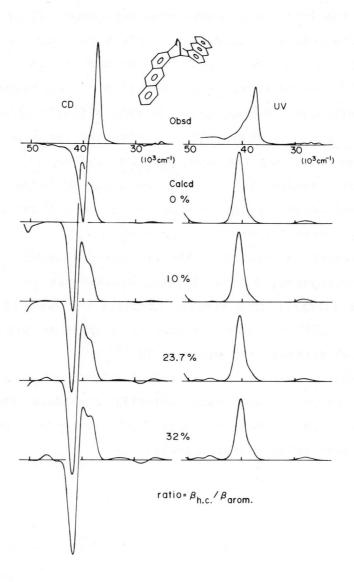

Figure 12-7. *The CD and UV spectra of (6R,15R)-(+)-4 obtained by the SCF-CI-DV molecular orbital calculation, in which the value of interchromophoric homoconjugation integral was changed. [Reprinted from reference 7.]*

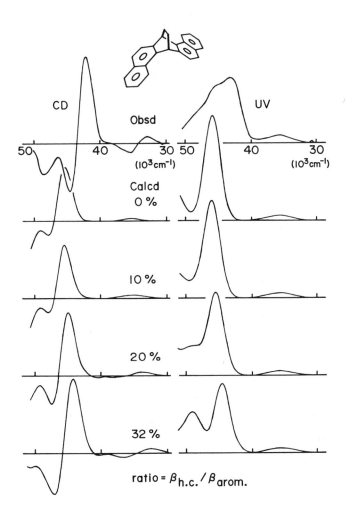

Figure 12-8.   *The CD and UV spectra of (7R,14R)-(+)-5 obtained by the SCF-CI-DV molecular orbital calculation, in which the value of interchromophoric homoconjugation resonance integral was changed.   [Reprinted from reference 7.]*

References

1. D. Bohm, Quantum Theory (New York: Prentice-Hall, 1951), p. 427.

2. H. Eyring, J. Walter, and G. E. Kimball, Quantum Chemistry (New York: Wiley, 1944), Chapter 3.

3. A. Moscowitz, Tetrahedron 13, 48(1961).

4. C. M. Kemp and S. F. Mason, Tetrahedron 22, 629 (1966).

5. M. Kral, Collect. Czech. Chem. Commun. 35, 1939 (1970).

6. R. S. Mulliken, C. A. Rieke, D. Orloff, and H. Orloff, J. Chem. Phys. 17, 1248 (1949).

7. N. Harada, Y. Tamai, and H. Uda, J. Am. Chem. Soc. 102, 506 (1980).

8. N. Harada, Y. Takuma, and H. Uda, J. Am. Chem. Soc. 100, 4029 (1978); see also reference 9.

9. N. Sakabe, K. Sakabe, K. Ozeki-Minakata, and J. Tanaka, Acta Crystallogr. Sect. B 28, 3441 (1972).

10. N. Harada, to be published.

APPENDIX

I. Greek Alphabet

α	A	Alpha	ν	N	Nu
β	B	Beta	ξ	Ξ	Xi
γ	Γ	Gamma	o	O	Omicron
δ	Δ	Delta	π	Π	Pi
ϵ	E	Epsilon	ρ	P	Rho
ζ	Z	Zeta	σ	Σ	Sigma
η	H	Eta	τ	T	Tau
θ	Θ	Theta	υ	Υ	Upsilon
ι	I	Iota	ϕ, φ	Φ	Phi
κ	K	Kappa	χ	X	Chi
λ	Λ	Lambda	ψ	Ψ	Psi
μ	M	Mu	ω	Ω	Omega

II. The Theory of Optical Rotatory Power[1]

The right circularly polarized light propagated along the +z axis is formulated as

$$\vec{E} = E\ (\vec{i}\ \cos\phi - \vec{j}\ \sin\phi) \tag{1}$$

$$\phi = 2\pi\nu(t - \frac{nz}{c}) \tag{2}$$

$$v = \frac{c}{n} \tag{3}$$

where $\vec{E}$ is electric field vector, E is amplitude of $\vec{E}$, $\vec{i}$ and $\vec{j}$ are unit vectors of x and y axes, respectively, ϕ is phase of the wave, ν is frequency, t is time, n is refractive index, c is velocity of light in vacuum, z is z axis coordinate, and v is velocity of the light in the medium.

As shown in Figure 1, at t = 0 and z = 0, $\vec{E}$ is directed along the +x axis. At z = 0, as t increases, $\vec{E}$ rotates toward the -y axis. Namely, as time goes on, an observer facing the propagated light observes a clockwise rotation of $\vec{E}$ vector (Figure 1). If t = 0 and z is changed as z = $k\lambda/12$, (k = 0, 1, 2, 3,$\cdots$), equation (1) gives results as depicted in Figure 1. Namely, right circularly polarized light is composed of right handed helicity of E vector; (In some textbooks, it is mistakenly described that right circularly polarized light has left handed helicity).

When the right circularly polarized light enters an optically active medium, the refractive index for right circularly polarized light n_R is obtained as follows, by calculation of the Maxwell's and quantum mechanical equations (for further details see reference 1).

$$n_R = \varepsilon^{1/2} - 2\pi\nu g \tag{4}$$

$$g = 4\pi N_1 \left(\frac{\beta}{c}\right) \left(\frac{\varepsilon+2}{3}\right) \tag{5}$$

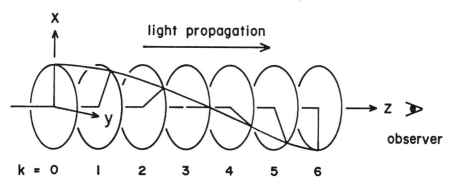

right circularly polarized light

Figure 1. Three dimensional structure of right circularly polarized light; $t = 0$ and z is changed as $z = k\lambda/12$, ($k = 0, 1, 2, \cdots\cdots$). Namely, right circularly polarized light is composed of right handed helicity of electric field vector. On the other hand, as time goes on, an observer facing the propagated light encounters the electric field vector in the order of 6, 5, 4, 3, 2, 1, 0, because the light travels toward him. Therefore he observes a clockwise rotation of the electric field vector.

$$\beta = \frac{c}{3\pi h} \sum_a \frac{Im\{<0|\vec{\mu}|a>\cdot<a|\vec{M}|0>\}}{\nu_a^2 - \nu^2} \qquad (6)$$

$$\nu_a = E_a/h \qquad (7)$$

where N_1 is number of molecules/cm^3, ε is dielectric constant, Im stands for imaginary part of the term in brackets, $<0|\vec{\mu}|a>$ and $<a|\vec{M}|0>$ are electric and magnetic transition moments of transition $0 \to a$ respectively, E_a is excitation energy, and h is Planck's constant.

The left circularly polarized light composed of left handed helicity as shown in Figure 2 is formulated as

$$\vec{E} = E (\vec{i} \cos\phi + \vec{j} \sin\phi) \qquad (8)$$

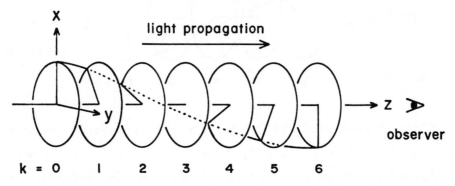

Figure 2. *Three dimensional structure of left circularly polarized light (see caption of Figure 1).*

In this case, the refractive index of the left circularly polarized light is calculated as

$$n_L = \varepsilon^{1/2} + 2\pi\nu g \tag{9}$$

Substitution of equations (4) and (9) into (1) and (8), respectively, gives

$$\phi_R = \phi_0 + \alpha \tag{10}$$

$$\phi_L = \phi_0 - \alpha \tag{11}$$

$$\phi_0 = 2\pi\nu(t - \frac{nz}{c}) \tag{12}$$

$$\alpha = 4\pi^2\nu^2 g \frac{z}{c} \tag{13}$$

where ϕ_0 is the phase of a wave propagated with the mean refractive index $n = \varepsilon^{1/2} = (1/2)(n_R + n_L)$.

From these equations,

$$\alpha = (\phi_R - \phi_L)/2 \tag{14}$$

$$\phi_R = 2\pi\nu(t - \frac{n_R z}{c}) \tag{15}$$

$$\phi_L = 2\pi\nu(t - \frac{n_L z}{c}) \tag{16}$$

$$\phi_R - \phi_L = \frac{2\pi \nu z}{c}(n_L - n_R) \tag{17}$$

$$\alpha = \frac{\pi z}{\lambda}(n_L - n_R) \tag{18}$$

The superposition of right and left circularly polarized lights of equations (1) and (8) gives a plane polarized light, as follows:

$$\vec{E} = E\{\vec{i}\ \cos(\phi_0 + \alpha) - \vec{j}\ \sin(\phi_0 + \alpha) + \vec{i}\ \cos(\phi_0 - \alpha) + \vec{j}\ \sin(\phi_0 - \alpha)\}$$

$$= 2E\ \cos\phi_0\{\vec{i}\ \cos\alpha - \vec{j}\ \sin\alpha\} \tag{19}$$

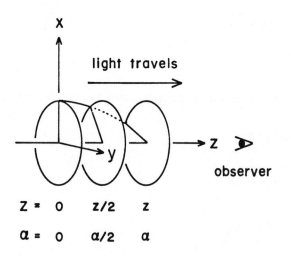

Figure 3. *Optical rotation in a dextrorotatory medium; an observer facing the propagated light observes a clockwise rotation of the polarized plane. The trace of the rotation constitutes a left handed helicity.*

If $\alpha = 0$, $\vec{E}$ is along the x axis; no optical rotation is observed. If $\alpha > 0$, $\vec{E}$ has been rotated by an angle α in a clockwise direction as viewed by an observer facing the direction of propagated light. Therefore, a medium with positive α is dextrorotatory (Figure 3). On the other hand, if $\alpha < 0$, $\vec{E}$ has been rotated in a counterclockwise direction; the medium is levorotatory.

The rotation in radian per unit length (cm) is formulated as

$$\alpha' = \alpha/z = \frac{4\pi^2 \nu^2 g}{c} = \frac{\pi}{\lambda}(n_L - n_R) \qquad (20)$$

Substitution of equations (5) and (6) into (20) gives

$$\alpha' = \frac{16\pi^3 N_1 \nu^2}{c^2} \left(\frac{n^2+2}{3}\right) \beta$$

$$= \frac{16\pi^2 N_1}{3hc} \left(\frac{n^2+2}{3}\right) \sum_a \frac{\nu^2 R_{a0}}{\nu_a^2 - \nu^2} \qquad (21)$$

$$R_{a0} = \mathrm{Im}\{<0|\vec{\mu}|a> \cdot <a|\vec{M}|0>\} \qquad (22)$$

Equation (22) is called Rosenfeld's equation, and R_{a0} is the rotational strength of the excitation from ground state 0 to excited state a. Thus, the rotational strength which governs the sign and amplitude of optical rotation is equal to the imaginary part of the scalar product between electric and magnetic transition moment vectors.

III. Recent Applications Added in Proofs.

1. Use of thiobenzoate chromophore and absolute configurational studies of asymmetric 1,4-addition of thiocarboxylic acids to 2-cyclohexenones (Table 1).[2]

Table 1.

(+)

(−)

CD (2)

cyclohexane

2. Absolute configuration of brevetoxin B isolated from the red tide Dinoflagellate Ptychodiscus brevis (Gymnodinium breve) (Table 2).[3]

Table 2.

244.5 (37,600)

253.0 (+14.3)

239.0 (−5.7)

CD(3)

MeOH

MeOH

3. A Micromethod for Determining the Branching Points in Oligosaccharides (Table 3).[4]

Table 3.

253 (−4)

238 (+2)

CD (4)

253 (+40)

236 (−18)

CD (4)

4. Chiral Molecular Association of Flower Color Pigments (Table 4).[5]

Table 4. Visible and CD Spectra of Anthocyanidin 3,5-Diglucosides $(5 \times 10^{-4} \text{M})$ in 0.1 M Phosphate Buffer at pH 7.0. [Adapted from reference 5.]

Anthocyanin	Visible, λ_{max} (nm), (ε)	CD, λ_{ext} (nm), ($\Delta\varepsilon$)	
		First Cotton	Second Cotton
Pelargonin	560 (16,400)	630 (+60.6)	550 (−20.0)
Cyanin		634 (+13.9)	520 (−10.9)
Cyanin[a]	586 (12,000)	630 (+78.8)	515 (−76.4)
Peonin	550 (17,000)	630 (−3.6)	525 (+3.0)
Delphin	544 (19,000)	600 (−6.7)	510 (+11.2)
Hirsutin	528 (8,900)	575 (−1.8)	490 (+1.8)
Malvin	564 (19,200)	650 (−3.6)	516 (+5.5)

[a] Aggregated precipitates obtained from aqueous solution at pH 7.0.

5. Absolute Configuration of Floridenol and Related Compounds, new sester-terpenes Isolated from a Pest Insect, C. floridensis (japonicus) Comstock (Table 5).[6]

Table 5.

	244	242 (+12)	CD (6)
	isooctane	isooctane	

6. Absolute Configuration of the Synthetic Intermediate for Sirodesmin A, a Fungal Metabolite (Table 6).[7]

Table 6.

(+)   CD (7)

7. Absolute Configuration of the Diol Metabolite of Precocene I (Table 7).[8]

Table 7.

(3R,4S)

325 (+52.9)[a]
312 (0.0)   CD (8)
302 (-28.6)

MeOH

[a] Estimated from CD curves.

8. Absolute Stereochemistry of Dihydrodiol Metabolites of Polyaromatic Hydrocarbons (Table 8).

Table 8.

(2S)

233 (-30.2)
228 (0.0)   CD (9)
222 (+24.4)

10% dioxane/
  MeOH

(3S,4R)

231 (+65.3)
226 (0.0)   CD (9)
220 (-30.7)

10% dioxane/
  MeOH

(5S,6S)

324 (+17.3)   CD (10)
285 (-4.5)

  MeOH

11. CD and UV Spectra of 1,3-Dibenzoate Systems (Table 11).[14]

Table 11.

302.6 (51,200)

228.0 (14,200)

EtOH

322.0 (-21.6)	
308.7 (0.0)	
296.0 (+13.9)	CD (14)
237.5 (-0.9)	
219.5 (+1.6)	

EtOH

312.0 (60,300)

227.8 (13,800)

EtOH

321.1 (-40.7)	
308.1 (0.0)	
295.5 (+19.7)	CD (14)
228.5 (-1.7)	

EtOH

The UV spectrum of the cis-isomer (1,3-diaxial dibenzoate) exhibits a remarkable blue shift of λ_{max}, in comparison with that of a monobenzoate (λ_{max} 311.0 nm), because the angle between two electric transition moments is approximately zero (see section 3-1-E). Since the CD spectrum shows negative first and positive second Cotton effects of moderate intensity, the two benzoate groups deviate to some extent from ideal diaxial position, constituting a negative exciton chirality.

12. CD Data of Some Monoterpene Allylic Alcohol p-Nitrobenzoates (Table 12).

Table 12.

Structure			
	258.5 (14,200) EtOH	*262.0 (+5.84)* EtOH	CD (15)
	258.9 (13,800) EtOH	*261.0 (-5.64)* EtOH	CD (15)
	258.4 (14,100) EtOH	*258.0 (+7.39)* EtOH	CD (15)
	258.8 (13,700) EtOH	*264.4 (+5.11)*[a] EtOH	CD (15)
	259.1 (14,300) EtOH	*262.0 (-5.45)*[a] EtOH	CD (15)

[a] The $\Delta\varepsilon$ values were corrected on the basis of the optical purity (72.6%).

Reference

1. H. Eyring, J. Walter, and G. E. Kimball, <u>Quantum Chemistry</u> (New York: John Wiley, 1944), Chapter 17. D. J. Caldwell and H. Eyring, <u>The Theory of Optical Activity</u> (New York: Wiley-Interscience, 1971). E. Charney, <u>The Molecular Basis of Optical Activity, Optical Rotatory Dispersion and Circular Dichroism</u> (New York: Wiley, 1979).

2. J. Gawronski, K. Gawronska, and H. Wynberg, <u>J. Chem. Soc., Chem. Commun.</u> 1981, p. 307.

3. Y. Y. Lin, M. Risk, S. M. Ray, D. V. Engen, J. Clardy, J. Golik, J. C. James, and K. Nakanishi, <u>J. Am. Chem. Soc.</u> 103, 6773 (1981).

4. H. -W. Liu and K. Nakanishi, <u>J. Am. Chem. Soc.</u> 103, 7005 (1981).

5. T. Hoshino, U. Matsumoto, N. Harada, and T. Goto, <u>Tetrahedron Lett.</u> 22, 3621 (1981).

6. Y. Naya, K. Yoshihara, T. Iwashita, H. Komura, K. Nakanishi, and Y. Hata, <u>J. Am. Chem. Soc.</u> 103, 7009 (1981).

7. W. H. Rastetter, J. Adams, and J. Bordner, <u>Tetrahedron Lett.,</u> 23, 1319 (1982).

8. R. C. Jennings, <u>Tetrahedron Lett.,</u> 23, 2693 (1982).

9.   M. Koreeda, M. N. Akhtar, D. R. Boyd, J. D. Neill, D. T. Gibson, and D. M. Jerina, J. Org. Chem., 43, 1023 (1978).

10. D. R. Thakker, W. Levin, H. Yagi, S. Turujman, D. Kapadia, A. H. Conney, and D. M. Jerina, Chem.-Biol. Interactions, 27, 145 (1979).

11. B. Kedzierski, D. R. Thakker, R. N. Armstrong, and D. M. Jerina, Tetrahedron Lett., 22, 405 (1981).

12. H. Yagi, K. P. Vyas, M. Tada, D. R. Thakker, and D. M. Jerina, J. Org. Chem., 47, 1110 (1982).

13. G. Bellucci, G. Berti, R. Bianchini, P. Cetera, and E. Mastrorilli, J. Org. Chem., 47, 3105 (1982).

14. N. Harada, unpublished data.

15. N. Harada, J. Iwabuchi, Y. Yokota, H. Uda, and K. Nakanishi, J. Am. Chem. Soc., 103, 5590 (1981).

INDEX

455